COMPREHENSIVE BIOCHEMISTRY

ELSEVIER PUBLISHING COMPANY

335 Jan van Galenstraat, P.O. Box 211, Amsterdam, The Netherlands

ELSEVIER PUBLISHING COMPANY LIMITED

Barking, Essex, England

AMERICAN ELSEVIER PUBLISHING COMPANY, INC.

52 Vanderbilt Avenue, New York, N.Y. 10017

Library of Congress Card Number 62–10359
ISBN 0-444-40870-X

With 44 illustrations and 32 tables

COMPREHENSIVE BIOCHEMISTRY

COMPREHENSIVE BIOCHEMISTRY

SECTION I (VOLUMES 1–4)
PHYSICO-CHEMICAL AND ORGANIC ASPECTS
OF BIOCHEMISTRY

SECTION II (VOLUMES 5–11)
CHEMISTRY OF BIOLOGICAL COMPOUNDS

SECTION III (VOLUMES 12–16)
BIOCHEMICAL REACTION MECHANISMS

SECTION IV (VOLUMES 17–21)
METABOLISM

SECTION V (VOLUMES 22–29)
CHEMICAL BIOLOGY

HISTORY OF BIOCHEMISTRY (VOLUME 30)

GENERAL INDEX (VOLUME 31)

COMPREHENSIVE
BIOCHEMISTRY

EDITED BY

MARCEL FLORKIN

Professor of Biochemistry, University of Liège (Belgium)

AND

ELMER H. STOTZ

Professor of Biochemistry, University of Rochester, School of Medicine and Dentistry, Rochester, N.Y. (U.S.A.)

VOLUME 26 PART C

EXTRACELLULAR AND SUPPORTING STRUCTURES

(continued)

ELSEVIER PUBLISHING COMPANY

AMSTERDAM · LONDON · NEW YORK

1971

CONTRIBUTORS TO THIS VOLUME

S.O. ANDERSEN, Dr. PHIL. (Copenhagen)
Assistant Director of Research, Department of Zoology, University of Cambridge, Downing Street, Cambridge (Great Britain)

RENÉ DEFRETIN, B.Sc., M.Sc., D. Sc.
Professor of Marine Biology, Dean of the Faculty of Sciences, (Director of the Institut de Biologie Maritime et Régionale, Wimereux), Laboratoire de Zoologie, Faculté des Sciences de Lille, B.P. 36, 59-Annappes (France)

JOHN E. EASTOE, D.Sc., Ph.D.
Research Fellow in Biochemistry, Department of Dental Science, Royal College of Surgeons, Lincoln's Inn Fields, Londen, W.C.2A 3 PN (Great Britain)

CARL FRANZBLAU, D.Sc.
Associate Professor of Biochemistry, Department of Biochemistry, Boston University School of Medicine, 80 East Concord Street, Boston, Mass. 02118 (U.S.A.)

CHARLES JEUNIAUX
Professor of Zoology, Department of Morphology, Systematics and Animal Ecology, Zoological Institute, Quai Van Beneden 22, Liège (Belgique)

MARGARET JOPE, B.Sc., D. PHIL.
Department of Geology, Queen's University, Belfast BT7 1NN (Northern Ireland)

As this volume went to press, a chapter on *Bone* was not yet available for publication. It is planned to include such a chapter in a supplementary volume.

CONTENTS

VOLUME 26 PART C

EXTRACELLULAR AND SUPPORTING STRUCTURES

(continued)

Chapter IX. Chitinous Structures

by CHARLES JEUNIAUX

Chapter X. Resilin

by S. O. ANDERSEN

Chapter XI. Elastin

by CARL FRANZBLAU

Chapter XII. The Tubes of Polychaete Annelids

by RENÉ DEFRETIN

Chapter XIII. Constituents of Brachiopod Shells

by Margaret Jope

Chapter XIV. Dental Enamel

by JOHN E. EASTOE

COMPREHENSIVE BIOCHEMISTRY

Chapter IX

Chitinous Structures

CHARLES JEUNIAUX

Institute Ed. van Beneden, Morphology, Systematics and Animal Ecology, University of Liège (Belgium)

1. Introduction: the concept of chitinous structures

Among the organic molecules used extensively by living organisms in the elaboration of skeletal or cuticular structures, chitin is clearly one of the most important. Chitin has widespread distribution, it is quantitatively abundant in the biosphere, and it occurs in many different kinds of structures.

Since the discovery[1] in 1799 of an organic material in the arthropod cuticle which was particularly resistant to the usual chemical reagents, a material called chitin by Odier[2], the term *chitinous structure* has been commonly used to designate all types of organic structures exhibiting the physical properties of the arthropod cuticle, regardless of chemical composition. Terms such as "pseudochitin" or "chitinoid" were introduced to characterize structures which exhibited only some of the typical properties of true chitin, and entomologists often used the expression "chitinized" to describe the most sclerotized parts of the insect integument.

It is clear that the term *chitinous structure* should be restricted to those formations in which chitin itself is present and plays a structural role. It will be clear from later discussion that chitin does not exist as such in chitinous structures, but is a product of degradation of naturally occurring chitin–protein complexes, for which the term *native chitin* has been proposed[3]. Thus we can define a chitinous structure as a skeletal or cuticular structure which consists of chitin–protein complexes, that is to say glycoproteins or mucoproteins, the prosthetic group of which is entirely or principally chitin.

[595]

The chitin–protein complexes appear to provide the framework which support organic and mineral deposits, and the latter probably confer the special properties of any different type of chitinous structures. Studies on the chemistry and ultrastructure of these highly complex structures have only just begun, although X-ray diffraction work has provided a better knowledge of the chitin–protein framework. This chapter is designed to relate the chemical and architectural organization of the chitin–protein complexes to the general properties of chitinous structures, and to the role these structures play in living organisms and biological evolution.

2. Chemical composition and molecular structure of chitin

Chitin is a linear homoglycan[4] consisting of N-acetyl-D-glucosamine units (2-acetamido-2-deoxy-D-glucose), linked by glycosidic bonds in the $\beta(1-4)$ position. The structural repeating unit is *chitobiose*, the dimer of N-acetyl-D-glucosamine. The chemical constitution and properties of chitin have been summarized in this treatise (Vol. 5, p. 208 and pp. 266–270). More exhaustive reviews of chitin chemistry are also available[5-7]. Only certain aspects of chitin chemistry essential to further discussion will be considered here.

(a) Physical properties

Chitin is a chemically stable constituent of chitinous structures. Chitin is isolated only after destruction or removal of the other constituents. After removing mineral deposits and organic substances by drastic treatments from a typical chitinous structure, for example a lobster shell or an insect cuticle, sheets of "pure" chitin may be obtained. These sheets are colorless, or white in color if thick enough, and bear all the morphological characteristics of the starting shell or cuticle.

However, isolated chitin sheets do not exhibit the remarkable properties of the original chitinous structures, such as hardness, rigidity, or impermeability. Chitin sheets do show a relatively high tensile strength; according to the species, values are reported to be between 10 and 58 kg/mm^2 in the case of dry chitin sheets isolated from arthropod cuticles, and about 2 kg/mm^2 in the case of moist material[8-10]. Other physical properties of isolated chitin differ considerably from those exhibited by the naturally occurring chitinous structures. Sheets of isolated chitin are soft and pliable, and are permeable to water and gases. Clearly, many of the mechanical and physical properties of chitinous structures are not due to chitin itself, but to chitin complexes and deposits.

(b) *Nature of the amino-sugar residues in chitin*

Elucidation of the nature of interactions between chitin molecules and between chitin and other substances depends on precise knowledge of the chemical composition of the polysaccharide. While it is generally assumed that chitin is built only of N-acetyl-D-glucosamine units, the nitrogen content of purified chitin has been reported by numerous workers to be 6.1–6.7%, while the calculated value is 6.89%. These discrepancies are generally explained by incomplete purification of the chitin, or by the fact that chitin is a strong adsorbent, sometimes used indeed in chromatography.

Probably more significant are observations that the enzymatic hydrolysis of pure chitin by chitinase (EC 3.2.1.14) plus chitobiase (EC 3.2.1.29) leads to the production of small amounts of free glucosamine along with acetylglucosamine[11,12] or to a triose in which one or two amino groups are not acetylated[12]. The presence of glucosamine is apparently not due to the action of deacetylases, since such enzymes were not detected in the enzymatic extracts used[11,12]. There is a possibility that partial deacetylation occurs during chitin isolation.

These observations led Giles *et al.*[13] to a re-examination of the chemical composition of chitin extracted by mild procedures. The analysis of C, H, and N in different samples of purified and dried lobster chitin suggests that the proportion of amino sugars are 82.5% acetylglucosamine, 12.5% glucosamine, and 5% water. Chitin chains would thus be composed of about 1 glucosamine for 6–7 acetylglucosamine residues, in addition to firmly bound water. This interpretation has been discussed at length by Rudall[14], who concluded on the basis of X-ray-diffraction diagrams, infrared-absorption spectra and density measurements, that chitin may depart from an idealized poly-N-acetylglucosamine structure in having one residue deacetylated for every 6 or 7 residues, with bound water replacing missing acetyl groups to maintain density and crystallographic properties consistent with those of a poly-N-acetylglucosamine. A similar interpretation is given for the chitin of cuttlefish shell and squid "pen", which is of the crystallographic β-type[12].

(c) *Macromolecular structure*

The chain configuration of chitin has been difficult to elucidate, due to its insolubility and the possibility of structure modification during isolation. It is obvious however, from the observations of Herzog[15] and Gonell[16] that the structure is crystalline and shows a great similarity to that of cellulose. The studies of Meyer and coworkers[17,18] on crustacean chitin led to the classical figure of an orthorhombic unit cell, having the dimensions $a = 9.40$ Å, $b = 10.46$ Å, and $c = 19.25$ Å (Fig. 1). Chitin fibers are extended along the b-axis, and adjacent chains run in opposite directions. Owing to the β-glycosidic linkage between acetylglucosamine residues, the acetamido groups alternate from one side to the other along the chitin chain.

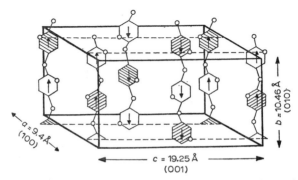

Fig. 1. Unit cell of chitin, after Meyer and Pankow[18]. Arrows indicate the alternating directions of chains.

Using more elaborate methods, Carlström[19] confirmed these general conclusions, but proposed a simpler orthorhombic unit cell, the value of a being only 4.76 Å. This would mean that the chitin unit cell is composed of two chains, antiparallel in direction. The study of infrared absorption spectra permitted the determination of the position of intramolecular and interchain bonding. According to Carlström[19] the adjacent residues of the same chain are not only covalently bonded by $\beta(1-4)$ glycosidic linkages, but also hydrogen bonded between O_3 and O_5'. The neighbouring chains are linked by hydrogen bonds $CO \cdot \cdot H \cdot N$ between adjacent aminoacetyl groups; presumably all the NH groups of a chitin chain are hydrogen-bonded to the CO groups of the adjacent chain. These linkages and the general structure of the chitin unit cell are presented in Fig. 2.

Comparable data have been obtained with chitins isolated from fungi and insects by several investigators[20-24], who generally agree with Carl-

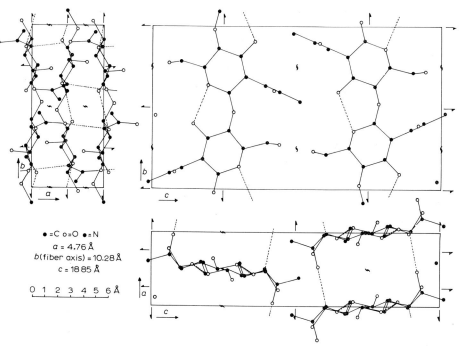

• =C o =O ●= N
a = 4.76 Å
b (fiber axis) = 10.28 Å
c = 18 85 Å

0 1 2 3 4 5 6 Å

Fig. 2. Structure of the unit cell of chitin (α-chitin), after Carlström[19]. Hydrogen bonds are shown as dotted lines.

ström's interpretation. Dweltz[25] however proposed another structure, derived from X-ray-diffraction studies of the lobster tendon. The dimensions of the orthorhombic unit cell was reported as $a = 4.69$ Å, $b = 10.43$ Å, and $c = 19.13$ Å (b being the fiber axis). These dimensions are similar to those earlier reported, but the positions of the hydrogen bonds are said to be different. As shown in Fig. 3 the NH and CO groups in the neighbouring aminoacetyl chains are hydrogen bonded, as are the hydroxymethyl side-chains containing a hydroxyl group. The short hydroxyl is intra-hydrogen bonded to the oxygen of the amide group in the same asymmetric unit, in such a way that the chain structure would be straight instead of buckled[25].

This modified view of the molecular organization of the chitin chains has been carefully examined by several investigators[14,26,27], who conclude that the interpretation of Carlström seems to be the most satisfactory, especially from a stereochemical viewpoint.

X-Ray-diffraction studies of diverse supporting structures indicate three

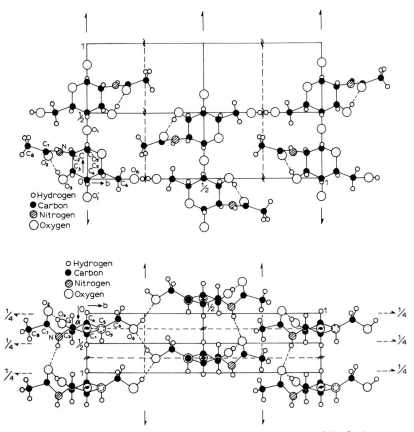

Fig. 3. Structure of α-chitin, after Dweltz[25]. Upper: the *a*-projection of the final structure.
Lower: the *c*-projection of one half of the unit cell.

different types of crystallographic patterns among chitins. These patterns
are presented in Fig. 4, according to Rudall[14]. The type of chitin discussed
above, commonly found in arthropods and fungi, has been named α-chitin
(Fig. 4B). A second type, β-chitin (Fig. 4A) has been discovered by Lotmar
and Picken[23] in the chaetae of the annelid polychaete *Aphrodite aculeata*,
and in the "pen" of the squid. The unit cell dimensions of isolated β-chitin
are $a = 9.32$ Å, $b = 10.17$ Å, and $c = 22.15$ Å. β-Chitin is more readily penetrated by chemical reagents and enzymes than α-chitin[12], suggesting a lower
degree of "packing" of chains and a more open type of crystalline structure.
Numerous free amino groups are said to exist within the chains, one or two

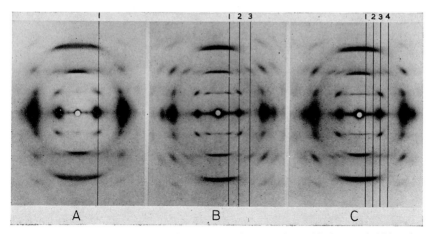

Fig. 4. X-Ray-diffraction patterns of the three main crystallographic forms of chitin, after Rudall[14]. A, β-chitin; B, α-chitin; C, γ-chitin. The essential differences between the diagrams are indicated by the approximate positions of the vertical row lines (1, 2, 3, etc.).

every five residues being unacetylated[12]. Nevertheless, after hydrolysis of β-chitin by chitinolytic enzymes, only small amounts of glucosamine occur, about 3 % of the quantity of acetylglucosamine[12].

When β-chitin is dissolved in formic or nitric acid and reprecipitated by dilution, the regenerated chitin exhibits the α-chitin pattern. According to Rudall[14] and Dweltz[28], water must occupy a significant part of the crystal structure of β-chitin, perhaps one molecule per acetylglucosamine residue. A triclinic model has been proposed for the unit cell of β-chitin[28].

A third type, γ-chitin, has been recently discovered by Rudall[14,29] in the thick cuticle lining the stomach of the squid *Loligo*. Its X-ray-diffraction pattern is seen in Fig. 4C.

Rudall[14,29] proposed an attractive interpretation of these three distinct crystallographic types of chitin, differing only in the number of chitin chains in the unit cell. Fig. 5 summarizes Rudall's interpretation. In the β-chitin, each unit cell contains only one chitin chain, the different chains running in parallel directions. In α-chitin there are two chains per unit cell, running in antiparallel directions. Such an arrangement might be obtained by folding each chain of β-chitin as seen in Fig. 5E. This is supported by the β to α transformation described above, which occurs with a contraction in length of about 50 %. Finally, because of the spacing of 29.2 Å in the plane of the sugar ring, given by the X-ray-diffraction diagram (Fig. 5C), the crystallites of

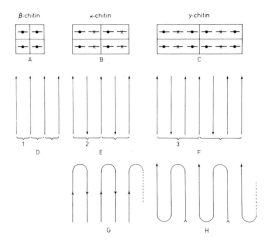

Fig. 5. Interpretation of the three main crystallographic patterns of chitin, after Rudall[14,29]. Diagrammatic representation of four unit cells of β-chitin (left), α-chitin (centre) and γ-chitin (right). In the first line, A, B and C represent basal projections of four unit cells, looking at the cell base along the length of the chains. The horizontal lines represent the plane of the sugar ring, and the signs ● and × represent two opposite directions with respect to the plane of the paper. In A (β-chitin), there is one chain per cell; in B (α-chitin), two chains per cell: in C (γ-chitin), three chains per cell. In the second line (D, E and F), chitin chains are viewed in the plane of the sugar rings and are arranged in groups of one, two or three chains. The third line represents the possible origin of the different chain orientations by folding every chain to give two (α-chitin) or three (γ-chitin) parallel segments (G and H).

γ-chitin must be formed by three chains, the central one running antiparallel between the two adjacent ones, a situation which could result from a folding of the β-chain with its extremities fixed in some way[29].

3. Distribution and chemical composition of chitinous structures

(a) Detection of chitin

Before considering the ultrastructural organization of chitinous structures, these structures should be classified. However, there is a great variety of supporting chitinous structures, especially in animals, about which little is known other than that they appear to contain chitin, usually associated with protein. The situation has been further complicated by the earlier lack of accurate analytical or histochemical methods for chitin.

The "chitosan" test is sometimes unreliable. The X-ray-diffraction method is more highly specific, at least when appreciable amounts of chitin are present, and Rudall[24] succeeded in revealing or confirming the presence of chitin in several structures. The use of an enzymatic method, based on the highly specific and purified chitinases[30,31], has led to a systematic study of the distribution of chitin in the animal kingdom. The principle of the method earlier suggested[32,33] has been developed by Jeuniaux[30,31] so as to give accurate quantitative data. Since the method depends on the amount of acetylglucosamine liberated after complete hydrolysis, the chitin value may be lower than the actual to the extent that some non-acetylated glucosamine exists in the chitin chains.

(b) Occurrence and chemical features of chitinous structures

The available data on chitin analyses are assembled in Table I, and lead to the following comments.

(1) Chitinous structures are very widely distributed, particularly in animals, and exist even in less evolved taxonomic groups in which chitin was earlier believed to be absent, as in Protozoa. The biosynthesis of chitin thus appears to be controlled by early established genes, present probably in the primitive unicellular root of Metazoa[30,31]. This biosynthetic ability has been retained by a number of diblastic animals and by most of the triblastic Protostomia, but was lost at the beginning of the deuterostomian evolutionary lineage[31], with the possible exception of Tunicata, the peritrophic membrane of which is said to contain chitin[81,82]. Echinoderma, Enteropneusta, Pterobranchia, Urochordata and Vertebrata have utilized other polysaccharides such as cellulose and chondroitin sulphates, or fibrous proteins such as collagen and keratins, for their supporting structures. In plants, chitinous cell walls or structural membranes are only found in those forms, such as moulds and fungi, which like animals find considerable combined nitrogen in their food sources. In contrast, photosynthetic plants utilize nitrogen-free sugars almost exclusively for their supporting structures; chitin is however said to be a constituent of the cell membrane of some lower green plants (Chlorophycea)[40].

(2) Chitinous structures are mainly, if not exclusively, of ectodermal origin in pluricellular animals[24,30]; thus they form the characteristic exoskeletons of most of the "invertebrates". This is in contrast with collagenous structures which are almost entirely of mesodermal origin[24].

(3) Chitin rarely constitutes more than 50% of the total organic matter in chitinous structures. Higher concentrations (up to 85%) are found only in

Organisms	Structures	Principal types of mineralizing substances
Fungi: Ascomyceta[b] Basidiomyceta Phycomyceta[c] Imperfecti(Moniliales)	cell walls and structural membranes of mycelia, stalks and spores	—
Algae: Chlorophyceae	cell wall	—
Protozoa: Rhizopoda (*Pelomyxa*)	cyst wall	—
shelled Rhizopoda (Plagiopyxidae)	shell	silica
shelled Rhizopoda (*Allogromia*)	shell	iron
Ciliata	cyst wall	—
Cnidaria: Hydrozoa		—
Hydroïdea	perisarc	
Milleporina	coenosteum	$CaCO_3$
Siphonophora	pneumatophore	—
Anthozoa (*Pocillopora*)	"skeleton"	$CaCO_3$
Scyphozoa (*Aurelia*)	podocyst	—
Aschelminthes: Rotifera	egg envelope (inner membrane)	—
Nematoda	egg capsule (middle membrane)	—
Acanthocephala	egg capsule	—
Priapulida	cuticle	—
Endoprocta	cuticle	—
Bryozoa (Ectoprocta)	ectocyst	sometimes $CaCO_3$
Phoronida	tubes	—
Brachiopoda: Articulata	stalk cuticle	—
Inarticulata (*Lingula*)	stalk cuticle	—
	shell	$CaCO_3$
Echiurida	hooked chaetae	
Annelida: Polychaeta	chaetae	—
Polychaeta (Eunicidae)	jaws	unidentified
Oligochaeta	chaetae; gizzard cuticle	—
all	peritrophic membrane	—

I

DIVERSE CHITINOUS STRUCTURES IN LIVING ORGANISMS

Chitin		Main other organic constituents (figures as % organic fraction)	References
% organic fraction (dry weight)	Crystalline type[a]		
traces to 45.0[d]		polysaccharides such as glucans or mannans	33–38
+ ?[e]	—	cellulose	40
+	—	—	41
+	—	—	30
+	—	unidentified proteins and lipids	30, 43
+	—	unidentified proteins	original data
3.2–30.3	$n(\alpha?)$	unidentified proteins, sometimes tanned	24, 30
+	$n(\alpha)?$	—	24, 44
+	$n(\alpha ?)$	—	24, 45
+	n	unidentified proteins	44, 46
+	—	unidentified proteins	47, 48
14.6[f]	—	unidentified proteins	49
16.6[f]	n	unidentified proteins	30, 50, 51
+	—	—	52
+	—	tanned proteins	53
+	—	tanned proteins	54
1.6–6.4	—	unidentified proteins	30, 55, 56
13.5	—	unidentified proteins	30
3.8	—	—	30
+	γ	collagen	24, 29
29.0	β	—	24, 30
+	—	—	30
20.0–38.0	β	quinone-tanned proteins	23, 30, 57
0.28	—	unidentified proteins	58
+	β	—	24
+	—	proteins	59

References p. 629

TABLE I

Organisms	Structures	Principal types of mineralizing substances
Mollusca: Polyplacophora	shell plates; mantle bristles	CaCO$_3$
	radula	iron
Gastropoda	shell (mother of pearl)	CaCO$_3$
	radula	iron and silica
	jaws	—
	"stomacal plates" (Opisthobranchia)	—
Cephalopoda	calcified shell	CaCO$_3$
	"pen" (*Loligo, Octopus*)	—
	jaws and radula	—
	stomach cuticle	—
Lamellibranchia	shells { periostracum	sometimes CaCO$_3$
	prisms	CaCO$_3$
	mother of pearl	CaCO$_3$
	calcitostracum	CaCO$_3$
	gastric shield	—
Onychophora	cuticle	—
Arthropoda: Crustacea } Diplopoda	calcified cuticle	CaCO$_3$
	intersegmental membranes	—
Insecta Arachnida } Chilopoda	hardened cuticle	—
	unhardened cuticle	—
all	peritrophic membrane	—
Chaetognatha	grasping spines	—
Pogonophora	tubes	—
Tunicata	peritrophic membrane	—

[a] The signs α, β and γ refer to the three crystallographic types defined by Rudall[14]; n means that the X-ray-diffraction pattern of chitin has been recorded, but not accurately identified as one of the former types.

[b] With the probable exception of *Saccharomyces* and related yeasts[32,33,39].

[c] With the possible exception of some Oomycetes and Monoblepharidiales[35,36].

(continued)

Chitin		Main other organic constituents *(figures as % organic fraction)*	References
% organic fraction (dry weight)	*Crystalline type*[a]		
12.0	—	unidentified proteins	30
+	—	unidentified proteins	60
3.0–7.0	—	conchiolin	30, 61
19.7	α	tanned proteins	24, 30, 62, 63
+	—	tanned proteins	24, 64
36.8	—	unidentified proteins	65, 66
3.5–26.0	β	conchiolin	24, 30, 67
17.9	β	"conchagen"	23, 30, 68
19.5	α	tanned proteins	24, 30
—	γ	—	29
0–7.3	—	unidentified proteins	30, 61
traces –0.2	—	conchiolin	30, 61
0.1–1.2	—	conchiolin	30, 61
0.2–8.3	—	conchiolin	30, 61
17.3	—	—	69 and original data
+		unidentified proteins	23, 24
58.0–85.0	α	arthropodins+sclerotins (10–32%)	24, 32, 70–72
48.0–80.0	α	arthropodins (23–51 %)	
20.0–60.0	α	arthropodins+sclerotins (40–76%)	22, 24, 32, 70, 71, 73–75
20.0–60.0	α	arthropodins+ (in some parts) resilin	
3.8–22.0	—	unidentified proteins (21–47%) + mucins	76, 77
+	—	—	55
33.0	β	unidentified proteins (47 %)	78–80
+[g]	—	—	81, 82

[d] Subject to considerable variation with age and culture conditions[37,42].

[e] Chitosan test, doubtful[73].

[f] Per cent of the total dry weight of the eggs.

[g] Identified by chitosan test only.

Arthropoda, which have exploited the aptitude of synthesizing chitin to a maximum. On the other hand, some calcified chitinous structures such as the ectocyst of Bryozoa and the shells of Gastropoda and Lamellibranchia, contain only small amounts of chitin. There are however no apparent relationships between the proportions of chitin and the degree of calcification, hardness, or flexibility of the structures.

(4) Chitin is associated with other polysaccharides in the cell walls of fungi. In animal forms chitin is associated with proteins, the latter frequently unidentified. In many instances of hard structures, the proteins are tanned by phenolic derivatives, namely the quinones. Collagen is rarely found in chitinous structures; the frequent independence between collagen secretion and chitin synthesis has been emphasized by Rudall[24].

(5) The distribution of the three different crystallographic forms of chitin does not seem to be related to taxonomy. Moreover, these three types may occur in different organs of the same animal, as in *Loligo* and *Lingula*[24,29]; it is possible that the three forms are associated with different functions[29]. According to Rudall[24], β-chitin (and probably γ-chitin) appears to be associated with collagen-type cuticles, or with collagen-secreting neighbouring tissues, while α-chitin structure completely replaces collagen-type cuticles.

(6) The various chitinous supporting structures exhibit striking differences in morphology, chemical composition and physical properties. The better defined chitinous structures can be classified as follows.

(i) *Flexible chitinous structures*

A typical flexible chitinous structure is the procuticle of the intersegmental membranes of arthropods. The cuticular sclerites also have the same characteristics just after moulting, *i.e.* before hardening by calcification or sclerotization. The chitinous procuticle is laid down by the epidermal cell layer, and it shows internal laminations visible with the light microscope. It is perforated by numerous minute ducts, the dermal ducts and the pore canals[84-88], through which the secretions of the epidermis and of the tegumental and dermal glands can reach the outside and form an outer non-chitinous layer, the epicuticle. In insects, the wax layer of the epicuticle is responsible for the water-proofing properties of the cuticle (for a detailed description of the epicuticle, see Locke[89]). The bulk of the procuticle is a chitin–protein complex, the protein fraction consisting of "arthropodins", which are not cross-linked by tanning reactions.

The innermost thin layer of the calcified cuticle of crustaceans ("membranous layer") is non-calcified; it is often referred to as endocuticle because it lies between the calcified procuticle and the epidermis, and possesses the same characteristics as the intersegmental membranes.

The peritrophic membrane of arthropods, and some other invertebrate groups, which wrap the aliments in the mesenteron and the feces in the postintestine, is an extremely thin, pliable, and hygroscopic membrane, the ultrastructure of which is described below.

A particular type of flexible chitinous structure has been recently identified in the wing hinges and elastic thoracic ligaments of the insect flight system [74,75]. Such cuticles are characterized by the presence of a rubber-like protein named resilin[90].

(ii) Sclerotized chitinous structures

The sclerites of insects and arachnids differ from intersegmental membranes by the fact that the outer layers of the procuticle are hardened by the quinone tanning of proteins. The resulting tanned proteins are the sclerotins. The outer sclerotized layers of the procuticle, adjacent to the non-chitinous epicuticle, constitute the exocuticle, while the innermost non-sclerotized layers constitute the endocuticle. There is no significant difference in the total amount of proteins between exocuticle and endocuticle of sclerites and procuticle of intersegmental membranes, but tanned scleroproteins are only found in the former layer.

The hardening of the exocuticle is often accompanied by a general darkening, i.e. by the formation and deposition of melanins. It has been demonstrated however that the processes of hardening and of darkening can occur separately[91,92]. It seems that melanization and sclerotization are catalysed by two different phenol oxidases[91,93].

In the hardest insect sclerites, such as in Coleoptera and Hymenoptera, the exocuticle is very thick, occupying almost the whole thickness of the former procuticle. In contrast, the sclerotized exocuticle forms only a very thin layer in the integument of insect larvae, mainly built by the endocuticle.

Histochemical observations reveal the existence, in some species, of a layer intermediate between endocuticle and exocuticle, named mesocuticle, which possesses peculiar staining properties[94]. The mesocuticle is particularly apparent and thick in arachnids[95].

In other hard chitinous structures, such as the cuticle (periderm) of Hydrozoan polyps and of Endoprocta[54], the cuticle of Priapulida[53], the chaetae

610 CHITINOUS STRUCTURES IX

of Annelida[57], and the jaws and radula of gastropods[62], there is also evidence for the existence of quinone-tanned proteins.

(iii) Calcified chitinous structures

One can consider the existence of two different types of calcified chitinous structures, in particular with respect to the proportions of chitin: (*1*) the sclerites of crustaceans and diplopods, in which chitin amounts to more than 50 % of the organic matter, and (*2*) the shells of molluscs (except cephalopods), in which chitin generally amounts to less than 10 % of the organic matter, *i.e.* only less than 0.4 % of the total weight of the shell.

In molluscan shells, chitin is often associated with a fibrous protein, conchiolin, in which sclerotization by quinone tanning seems to be entirely lacking. The calcified molluscan shells are fully described elsewhere in this treatise (see Vol. 26A, Chapter IV). The chemical composition of the shells is more variable in the cephalopods. In the external calcified shell of the Nautiloidea, the proportion of chitin is low, as in other molluscan shells. In the Decapoda, the shell is an internal skeletal structure, calcified in the Sepioidea, but non-calcified in the Theuthoidea ("pen" of the squids). This internal structure is considerably reduced in the Octopoda. In these non-calcified shells, the proportion of chitin is significantly higher (about 20 % of the total dry weight). According to Stegemann[68], the protein fraction, named "conchagen", of these internal structures can be converted to soluble proteins ("gelatins") by steam (143°); the remaining chitin is linked to a small protein moiety (2 % of the total residue).

The sclerites of crustaceans (at least Decapoda) can be described as intersegmental membranes whose procuticle is almost entirely calcified by calcium carbonate, and to a much lesser extent by calcium phosphates. Calcium carbonate occurs as micro or macro crystals of calcite or, in rare cases, of vaterite[96]. Calcification in the cuticles of Crustacea and Mollusca has been fully studied by Travis[97-100] and is discussed elsewhere in this treatise (see Vol. 26A, Chapter IV). The hardening of the calcified cuticles seems however to be initiated by protein sclerotization, prior to the deposition of calcium salts[101]; calcified crustacean cuticles indeed always contain definite amounts of sclerotins. As a result of the calcification, the amount of proteins in the calcified cuticle is significantly lower than in the flexible procuticle or in the sclerotized exocuticle of insects. It can be considered that mineral deposits replace a large proportion of the protein material in the hardening process[32].

(*iv*) *Non-calcified mineralized chitinous structures*

There are a limited number of cases in which hardness is provided by mineral deposits other than calcium salts. The shells of some shelled Rhizopoda (Thecamoebia) and the teeth of the radula of some mollusks are hardened by iron (Fe_2O_3) or silica (SiO_2) deposits. Despite some analytical and histochemical studies[60,62,63,102,103], our knowledge of the chemistry of these structures is too limited to allow further description.

In spite of the fact that the various types of chitinous structures exhibit striking physical and chemical differences, most of them show an association of chitin with protein. The problem of chitin–protein binding and the resulting ultrastructure are therefore more fully discussed below, as well as the biochemical processes of synthesis and degradation of these structures.

4. The chitin–protein complexes

Among the chitin–protein complexes, the cuticular proteins of arthropods have received particular attention. There are at least two different components, or mixtures of components having the same properties, (*1*) a water-soluble fraction, arthropodin[22], and (*2*) a water-insoluble fraction, sclerotin[104-105]. Arthropodins are soluble in hot water, but not in cold 10% trichloracetic acid. A suitable solvent for the extraction of arthropodins is that proposed by Trim[106,107]. Amino acid analyses have been performed by several authors on arthropodins of insects[106-110] and of crustaceans[111].

The arthropodins are characterized by the absence of the sulfur amino acids, the low proportion of glycine, and the high proportion of tyrosine. They thus seem to be essentially different from all other types of structural proteins[106,107]. A number of protein fractions with different solubility and electrophoretic properties have been isolated from the larval cuticle of the insect *Agrianome*, but their amino acid composition differs only slightly[109]. Despite their electrophoretic heterogeneity, the different arthropodin fractions of the larval cuticle of *Sarcophaga crassipalpis* behave in the ultracentrifuge as a monodisperse constituent, with a molecular weight of about 7000–8000[112].

The X-ray-diffraction pattern of arthropodins show the β-configuration, which is unusual among structural proteins. Fraenkel and Rudall[22] and Richards[32,113] pointed out that the molecular spacings of this protein in the extended configuration agree with those of chitin; the identity of these lattice spacings would presumably permit a mixed crystallization, and first suggested the possibility of a weak bonding between chitin and protein.

Experimental evidence of covalent bonds between chitin and protein in arthropods cuticles and in cephalopod shells has been obtained during the last decade. Acetylglucosamine as well as chitin can react with α-amino acids, especially tyrosine, peptides and cuticular proteins[114,115], to give stable complexes, dissociable, however, by changing pH values. The organic material of insect cuticles, decalcified crab cuticle and squid skeletal pen, dispersed in lithium thiocyanate, can be reprecipitated by acetone to give a series of chitin–protein fractions, presumably in the form of a glycoprotein complex[3], [109,116]. The proportion of protein in these complexes varies according to the methods of isolation and the material studied, from 7.5 % in the cuticle of *Cancer pagurus* to 51 % in the shell of the cuttlefish[3]. Finally, different samples of chitin, prepared by alkaline digestion of insect cuticle or decalcified crustacean cuticles, have been shown, in every case, to contain small amounts of aspartic acid and histidine[3]. In the case of the larval cuticle of the insect *Agrionome spinicollis*, there are, according to Hackman[3], two histidine and one aspartic residues per 400 residues of glucosamine.

It thus appears that chitin, whether in its α- or β-crystallographic form, is covalently linked to arthropodins or sclerotins to form more or less stable glycoprotein complexes, probably through aspartyl and histidyl residues. Owing to the relative stability of these complexes in hot alkali and their instability in hot acids, the linkage could probably be as an *N*-acylglucosamine[3], that is, between a carboxyl to the NH_2 group of glucosamine. Other covalent linkages, more labile in hot alkali, could probably occur in the chitin–protein complexes[14]. The protein components of the chitin–protein complexes of insect and crustacean cuticles show, according to Hackman[3], some differences in their amino acid composition, especially in glycine, lysine, and proline, but all of them contain significant amounts of aspartic acid and histidine. It must be noted that, in the internal shells of cephalopods, the greater part of the protein moiety ("conchagen") can be removed by hot water, whereas the remaining chitin is bound to a protein containing large amounts of aspartic acid[68].

These results however have not received confirmation by recent investigators[117], who did not succeed in the isolation of chitin–protein complexes, and did not find any predominance of aspartic acid and histidine, or any other amino acid, in the residual chitin after prolonged alkali treatments of *Calliphora* cuticles and *Loligo* pens.

According to Hackman and Goldberg[109] in the case of *Agrionome* larval cuticle, the protein fraction covalently linked to chitin amounts to 56 % of the total protein. In addition to this fraction there are 3 % of the total protein which is bound to other components by electrovalent bonds or double covalent bonds, 25 % linked by hydrogen bonds and 2 % bound by Van der Waals' forces. The remainder of the protein is not bound at all and is readily soluble in water. In certain parts of the cuticle of some insects ("rubber-like cuticles"), covalent linkages are said to occur between chitin and another type of protein, resilin[74,75].

The long radular ribbon of the Gastropoda and Polyplacophora provided Runham[60,62,63] with a convenient material to follow the sequence of events during the organization of the chitin–protein complexes. There is some histochemical evidence for the existence of covalent bonds between non-acetylated glucosamine residues of chitin and free carboxylic groups of protein, as in α-globulin and ovalbumin[118].

Whatever the exact nature of the chemical bonds between chitin and proteins, the stability of the chitin–protein complex is greatly enhanced by the sclerotization of the protein chains. This process consists of a polymerization of the polypeptide chains of "prosclerotin" by tanning with quinones. It is presumed that the prosclerotin is nothing but arthropodin. Due to sclerotization, the external part of the procuticle of an insect sclerite is transformed to exocuticle, the properties of which insure the hardness and the rigidity of the whole sclerite.

The quinone tanning of cuticular proteins is one of the physiological functions of the phenoloxidase system, which has been fully discussed elsewhere [119,120]. In the case of chitinous structures it must be emphasized that sclerotization proceeds outside the secreting epidermal cells, a rather short time after the deposition of the cuticular material in the form of a flexible chitinous structure. In arthropods, dihydroxyphenols diffuse from the epidermis through the procuticle by way of the pore canals, up to the external epicuticle where phenolases have been previously accumulated. After oxidation the resulting quinones diffuse back into the outer layers of the procuticle. The quinone reacts with a terminal amino group of a protein to form a N-catechol protein, which is then oxidised to a N-quinonoid protein. This compound reacts with the terminal amino group of an adjacent protein, forming a disubstituted derivative. Thus, a network of tanned proteins (sclerotins) is formed, including the chitin chains linked together by hydrogen bonds and covalently linked to the proteins. The degree of sclerotization determines the degree of hardness, rigidity and stability of the structure. The considerably higher stability given by sclerotization to chitinous structures is well illustrated by the resistance offered by the insect exocuticle to the powerful hydrolytic enzymes of the exuvial fluid during the molting processes (see below). In some cases, sulfur linkages are said to occur in the stabilization of the cuticular chitin–protein complex[121,122].

It is thus well established that the bulk of the chitinous structure is a glycoprotein in which chitin and protein are covalently linked, the proteins being, in sclerotized and many calcified structures, polymerized by a tanning process forming the chitin–sclerotin stable complexes. Beside this stable glycoprotein, there obviously exist free proteins, that can be easily extracted by water.

The study of the extent of chitin susceptibility to purified chitinases[123,124], before and after protein degradation by alkali treatment, provided a method to estimate the proportion of chitin bound or not bound to other substances in various chitinous structures[30,31]. It has been proposed to name "free chitin" that part of the chitin which is hydrolysed by pure chitinases in the intact structure, or eventually after decalcification. The "bound chitin" is only hydrolysed by chitinases after treatment with hot alkali. In most chitinous structures, the proportion of free chitin is low with respect to that of the bound chitin, the former amounting generally to only 4–30% of the total chitin. The annelid chaetae contain chitin almost entirely in the bound form, the free chitin amounting to only 0.5–2.2% of the total chitin. Two types of chitinous structures exhibit a quite different pattern; the free chitin is much more abundant in the shells of the molluscs (32–85% of the total chitin)[30,31] and in the peritrophic membranes of insects (25–68% of the total chitin)[31,77].

5. Ultrastructure

(a) Flexible chitinous structures

At the morphological level with the light microscope, the chitinous flexible structures such as insect endocuticle often appear with a lamellar organization. Numerous lamellae run parallel to the surface of the cuticle; their thickness may vary[32] from 0.2 to 10.0 μ. The lamellar structure is the result of a cyclic deposition of cuticular material by the epidermis[125–127]; morphogenesis of chitin lamellae can indeed be experimentally altered by varying light and temperature at the time of deposition[128]. Owing to the fact that, in cuticles, microfibers are only detected with the electron microscope after chem-

Fig. 6. Electron micrograph of lamellae and microfibres shown in a transverse section of the endocuticle of *Galleria* larva, from Locke[89]. Scale = 1 μ.

ical alteration, Richards[129,130] considered that such microfibres do not exist in normal insect cuticle. The observations of Locke[86,89] on intact insect endocuticles reveal however that lamellae do appear as being formed by sheets of microfibres (Fig. 6); these observations are confirmed by the findings of Neville[127,128]. The sheet arrangement of the microfibres corresponds to the dense part of each lamella, but the microfibres curve out at right angles between the sheets (Fig. 7). The word "lamina" has been proposed for

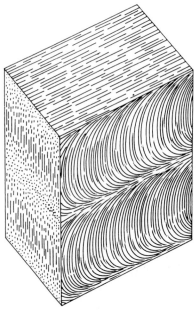

Fig. 7. Schematic representation of the arrangement of microfibres in the lamellae of a flexible chitinous structure, the endocuticle of an insect integument, from Locke[89].

that part of each lamella in which the microfibres run predominantly parallel to the surface[89]. The microfibrillar framework thus would occur in a three-dimensional pattern, an arrangement which is fundamentally different from that of plant cuticles[89]. Considering the results of autoradiographic experiments with labelled sugars and amino acids used as precursors of cuticular proteins and chitin[131], Locke[89] proposed the view that "chitin could be the molecule which is first ordered into microfibres".

A quite different interpretation has been recently proposed by Bouligand[132,133] who compares the lamellar ultrastructure to a "laminated wood", in the successive superposed planes of which the direction of the microfibres rotates regularly from the bottom to the top of the cuticle; such a geometrical organization could explain any type of fibrillar disposition like those shown in Figs. 6 and 7.

When measured on electron micrographs, the dimensions of the microfibres in the flexible chitinous structures of larval cuticles would be about 25 Å in diameter[134]. This order of size seems to agree satisfactorily with the data obtained by X-ray measurements for other caterpillar cuticles[14], namely 33 Å.

The ultrastructural association of proteins with chitin has been extensively studied by X-ray methods in a series of different chitinous structures and discussed at length by Rudall[14]. It appears that there exists a number of different types of associations, revealed by different types of altered or additional X-ray reflections; all the X-ray diffraction diagrams obtained so far are however "consistent with a structure in which protein fits exactly on the pattern made by small groups of chitin chains"[14].

The cuticle of the Onychophora is, with the exception of claws and jaws, a continuous flexible and unhardened chitinous structure, surmounted by a non-chitinous epicuticle. The ultrastructure of the procuticle of *Peripatopsis moseleyi*, examined by the electron microscope, has the general characteristics of that of the arthropod procuticle and endocuticle[135].

The microfibres constituting the fundamental pattern of flexible chitinous structures are much more obvious in such structures which do not require elaborate preparation and are thus not subjected to chemical alterations. This is the case of the ecdysial membrane of moth pupae[136], of the wing cuticula of *Ephestia*[45] and the tergal cuticula of *Lepisma*[45], of the cuticle of respiratory organs of arthropods[88], and of the peritrophic membranes of arthropods[137–140]. In the latter case the peritrophic membranes are formed by systems of strands arranged in a network pattern. The size and the geometrical disposition of the strands vary with the species considered; the strands are generally 0.10–0.20 μ, and are composed of a number of microfibrils, the diameter of which[123] is about 100 Å (Figs. 8 and 9). These microfibrils are sometimes embedded in a thin amorphous film. In many cases, three systems of fibrillar strands are placed at 60° to each other, delimiting hexagonal holes (Fig. 8). In other cases, there are only two systems of strands, disposed at right angles, delimiting more or less rectangular or lozenge-shaped holes[140, 141] (Fig. 9). The arrangement of the fibrils in three sets of strands is said to be

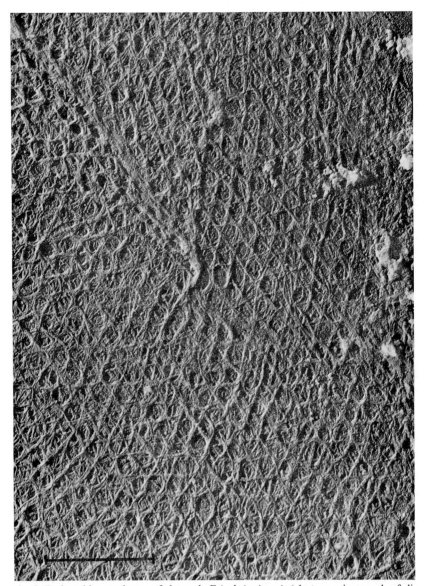

Fig. 8. Peritrophic membrane of the crab *Eriocheir sinensis* (electron micrograph of dissociated membranes deposited on to "Formvar" coated screen and shadowed with palladium; photograph Ch. Grégoire, unpublished). Scale = 1 μ.

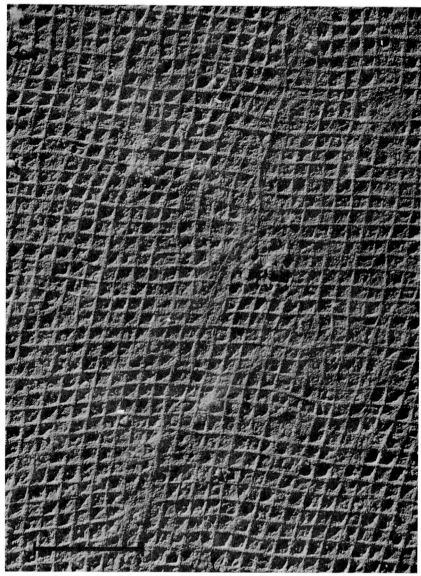

Fig. 9. Peritrophic membrane of the Diplopod *Julus albipes* (photograph Ch. Grégoire; legend and scale as for Fig. 8), after De Mets[141].

mechanically adapted to the task of forming a tough membrane not readily torn in any direction[139], but there seems to be no correlation between type of structure and either mode of formation of the membrane or nutrition of the species[140].

As already pointed out, chitin is present in the peritrophic membranes partly in a free state and partly in a bound state, probably in the form of glycoprotein complexes. The removal of the free chitin by pure chitinases does not alter the ultrastructural pattern, while removal of proteins by alkali causes a more pronounced dissociation of the strands into separate microfibres. Successive treatments by alkali and chitinase completely destroy the structure (Jeuniaux and De Mets, unpublished). It thus seems obvious that the chitin plays the fundamental role in the structural organization of the microfibres of the peritrophic membranes.

In the flexible tube of the pogonophores, chitin is highly crystalline and exhibits the β-crystallographic configuration. In this structure, oriented long ribbon-like fibrils can also be seen with the electron microscope, after removal of the proteins by alkali and dispersion by ultrasonic vibration[78]. These fibrils are about 1000 Å wide and 200 Å thick, this unusual large size being probably explained by the high crystallinity indicated by the X-ray-diffraction pattern[78].

(b) Sclerotized chitinous structures

The ultrastructure of chitin–protein systems in the hard exocuticles of insects is under extensive examination[14,142]. Combined studies of X-ray-diffraction patterns and electron micrographs will surely be successful in the near future in elucidating the exact ultrastructural features of the chitin–protein complexes in such cuticles. The electron micrograph obtained by Rudall[142] with the ovipositor walls of Hymenoptera (see Vol. 26B, Chapter VII, Fig. 1A, p. 563) is in close agreement with the X-ray-diffraction data, the low-angle diffraction patterns originating from a hexagonally packed system of chitin rods surrounded by uniform layers of proteins. The comparative microfibrillar organization in sclerotized chitinous structures and in keratins is considered by Rudall in the present treatise (Vol. 26B, Chapter VIII). The ultrastructure of the chaetae of Polychaeta, a very peculiar type of sclerotized chitinous structure, has been studied by Bouligand[143].

(c) Calcified chitinous structures

The chitin–protein complexes found in the molluscan shells probably also form microfibrils, dispersed in the well-known typical conchiolin structure extensively studied by Grégoire[144–147]. This conclusion can be drawn from the results obtained by Goffinet and Jeuniaux[148,149], in the case of the mother-of-pearl of *Nautilus* shell. After decalcification, the removal of the free chitin by purified chitinases does not modify the lace-like ultrastructure of the conchiolin (Fig. 10), while extraction of the insoluble nacrine (previously named nacrosclerotin[144]) of the conchiolin by mild alkaline treatments leaves an insoluble fibrous material, the "nacroine"[144]. The nacroine has

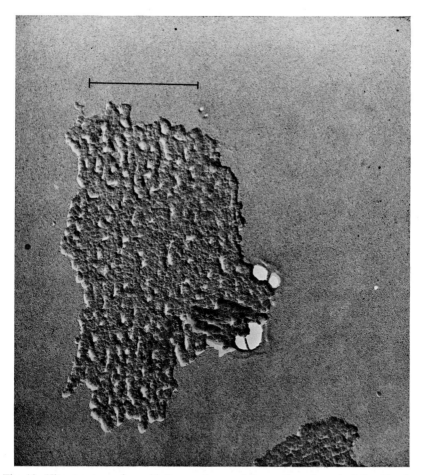

Fig. 10. Ultrastructure of conchiolin membrane of the mother-of-pearl of the *Nautilus* shell, after decalcification and treatment by pure chitinases. After Goffinet[149]. Scale = 1 μ.

been identified as a glycoprotein formed by the association of chitin and a polypeptide mainly built up of glycine and alanine[148]. This chitin–protein complex is shown with the electron microscope in the form of discrete microfibres apparently not arranged in a continous network[149] (Fig. 11). Microfibres of the same order of size had previously been observed by Grégoire [146,147] in the mother-of-pearl of lamellibranch shells, in the residue of

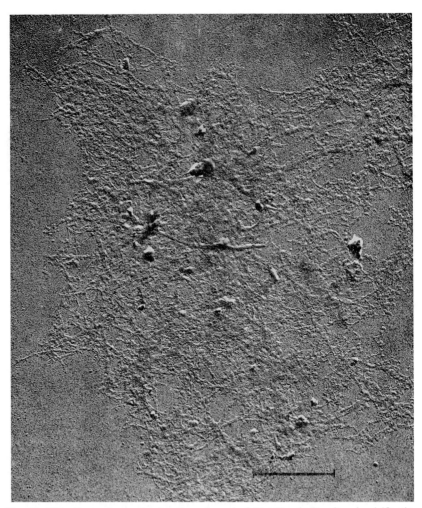

Fig. 11. Microfibres of nacroïne (mother-of-pearl of *Nautilus* shell), after decalcification and treatment by NaOH 0.5 N at 100° during 13 h. After Goffinet[149]. Scale = 1 μ.

decalcification of calcitic prisms of lamellibranchs, and in the porcelain layer of the shell walls of *Nautilus* (see also Florkin[150]). Following the detection of chitin in all these types of shell layers[30,61], it can be presumed that these fibrils could correspond to chitin–protein complexes.

The ultrastructure of calcified chitinous cuticles is of course mainly related to the problems of calcification.

(d) Concluding comments

As a general conclusion, it seems that the ultrastructural pattern of the different types of chitinous structures is dominated by the association of chitin chains with proteins, the chitin chains being grouped in sets of one, two or three, as described earlier. Despite the opinion of some authors claiming that microfibres are the result of chemical alteration, it appears that such microfibres are obvious in many cases. A microfibrillar organization is probably a general feature of all types of chitinous structures. As far as we know, it is presumable that chitin is the prime mover of this ultrastructural fibrillar pattern.

6. Synthesis and degradation of chitinous structures

(a) Formation of chitinous structures

There are very few data concerning the mechanisms of synthesis of chitinous structures, and a comprehensive description of this process is not at the present time within our reach. Chitinous structures are probably exclusively elaborated by cells of ectodermal origin[24,30]. In many cases, the activity of the secreting cells seems to be continuous throughout life, or at least until the building of the permanent structure is completed. This is for instance the case of the periderm of Hydrozoa, the ectocyst of Bryozoa, the jaws of molluscs and the chaetae of annelids.

The epidermis of arthropods is particular in that, owing to the existence of a series of moults during the life cycle, the secretion of the cuticle is a cyclic process controlled by the moulting hormone ecdysone. The coordination of events in the formation of the insect cuticle has been clearly described by Locke[89,151] from the point of view of a developmental biologist. In the present account, suffice it to say that, a short time before moulting, epidermal cells undergo a series of modifications. Accumulation of RNA is followed by a decrease at the time of moulting, numerous mitoses, followed by destruction of some of the cells, and puffing of the chromosomes. The epidermis retracts and secretes at first the moulting fluid (sometimes in the form of a gel[152]) which contains hydrolytic enzymes; the epidermis secretes also a thin membrane, the ecdysial membrane. The cuticulin layers of the epicuticle are laid down soon after, and the resorption of the hydrolytic products of the endocuticle takes place, a process of absorption which is obviously important but as yet poorly understood. Successive cuticular layers of procuticle, made up of chitin–protein complexes, are laid down beneath the cuticulin layers, partly before and partly after the shedding of the old cuticle (ecdysis). The new cuticle is expanded by absorption of water, as in crustaceans[153] and aquatic insect larvae, or by air swallowing and muscular efforts as in terrestrial insects [154–156]. When the whole procuticle is secreted and fully expanded, sclerotization takes place, together or not with calcification, in the future sclerite areas (for an extensive account of this process, see Cottrell[156]). The completed cuticle remains unchanged during the intermoult period till the next moult.

(b) Chitin biosynthesis

The biochemical process of chitin biosynthesis was first elucidated by Glaser and Brown[157] in the fungus *Neurospora crassa*. Acetylglucosamine units are transferred from uridine-diphosphate–*N*-acetylglucosamine (UDPAG) as donor on the end of a preformed chitodextrin chain used as a primer. This transfer is catalysed by a chitin–UDP acetylglucosaminyltransferase (EC 2.4.1.16). The same enzymatic system has been identified in different chitin-secreting organisms or tissues such as in cell-free homogenates of larvae, prepupae and pupae of the insect *Persectania eridania*[158], in homogenates of 3-days-old larvae of the crustacean *Artemia salina*[159] and in the epidermis of the crab *Callinectes sapidus* at the time of moulting[159]. The enzymatic system leading from glucose to UDPAG has been identified in the epidermis of the migratory locust[160]. In *Persectania*, the activity of the chitin–UDP acetylglucosaminyltransferase is the highest during the late final larval instar and the prepupal instar, *i.e.* exactly during the periods of maximal cuticle elaboration[161].

It thus appears highly probable that chitin biosynthesis is catalysed by the same enzymatic system in every type of chitinous structure, and that the genetic control of the synthesis of the enzyme chitin–UDP acetylglucosaminyltransferase is the prime mover of the fitness of an organism to build up chitinous structures.

(c) *Biological degradation of cuticular components during the life cycle of arthropods*

It has been emphasized[89] that only a small fraction of the arthropod chitinous exoskeleton is outside the body metabolic pool, in contrast to the exoskeletal and cuticular formations of other animals. When starved for a long time, some insects such as bugs and caterpillars are able to consume part of their cuticular components[89,162]. Degradation of the cuticular material occurs normally at every moulting period. The "cast skin" (exuvium) in only composed of the epicuticle and of the hardened exocuticle in the case of insects, while the ecdysial shell of crabs and lobsters only contains half of the chitin and one tenth of the arthropodins of the intermoult shell[30,163]. Moulting indeed involves the freedom of the epidermis from the old cuticle and the catabolism of a great part of the organic material constituting the cuticle; the two processes are realized due to the elaboration and the secretion of enzymes by the epidermis, namely proteolytic enzymes, chitinases and chitobiases, in the case of larval, pupal and imaginal insect moults[152,164,165], as well as in crustacean moults [166-169].

In *Platysamia cecropia*, the epidermis secretes a gel, enzymatically inactive for a time, which becomes active when it changes to a fluid state at the end of the pupal diapause[152]. However, in *Bombyx mori*, a moth without pupal diapause, the exuvial fluid is secreted immediately in its active form[30].

In *Bombyx mori*, as probably in other insects, the biosynthesis of chitinase by the epidermis is a cyclic process. The epidermal cells are devoid of chitinase during the intermoult period; the chitinolytic activity can only be detected a few hours before the secretion of the moulting fluid[30,170]. Chitobiases on the contrary are synthetized by the epidermal cells during the whole life cycle[30].

The processes are somewhat different in crustaceans, in which chitinase and chitobiase are elaborated throughout the intermoult period, but only secreted at the apical pole of the epidermal cell at the beginning of the stage D_1 (according to the terminology of Drach[153]). During the stage D_1, the non-calcified membranous layer undergoes hydrolysis of chitin and proteins, but there remains a very stable glycoprotein complex, containing chitin linked to proteins, which imbibes water and gels[168]; the enzymes of epidermal origin accumulate in this gelled layer, from which they diffuse outward[30] during the stages D_2 and D_3. These enzymes hydrolyse most of the chitin and the proteins of the calcified cuticle, with the exception of the chitin-tanned protein complexes[30].

References p. 629

In insects as in crustaceans, the degradation of the organic constituents of the cuticle is thus realized by the coupled action of two types of hydrolases: chitinases and proteolytic enzymes, which must act together in order to hydrolyse the chitin and protein molecules linked in the form of glycoprotein complexes[30].

(d) Degradation of chitinous structures in digestive processes

As emphasized above, the digestion of organic materials of chitinous structures requires the coupled action of chitinolytic and proteolytic enzymes. The adaptation of animals to a diet consisting of preys covered by chitinous cuticles or cell walls (fungi, zooplankton, arthropods), involves indeed the secretion of both types of enzymes by the glandular tissues of the digestive tract. Generally speaking, lack of chitin digestion is observed principally in animals which have adopted a chitin-free diet, such as herbivores. The distribution of chitinolytic enzymes and the correlation between the secretion of chitinase and the nature of the diet have been fully discussed by Jeuniaux[30].

(e) Degradation of chitinous structures in soils, waters and sediments

A wide number of bacteria, moulds and fungi are able to synthesize proteolytic and chitinolytic enzymes. These organisms, especially the streptomycetes in the soils, are responsible for the degradation of the chitinous structures of dead organisms. However, chitinous structures sometimes can be preserved during their burial in sediments and during geological periods. Chitin has indeed been identified in the wing-sheats of Coleoptera found in the eocene yellow amber[171] and even in a cambrian fossil pogonophore, *Hyolithellus*, which has withstood about 500 million years of fossilization[172].

7. Morphological radiation of chitinous structures

The chitinous structures, with the possible exception of those constructed by moulds and fungi, are built up by a glycoprotein framework, in which chitin is covalently linked to proteins. At this level of organization, the chitinous structures exhibit a high tensile strength but are essentially pliable and flexible, allowing movements and limited expansion. After sclerotization of the proteins, the resulting chitin-tanned protein complex gains considerable stability and confers to the structure hardness, rigidity and resistance to enzymatic hydrolysis. Moreover, these chitin–protein complexes can be completed by the deposition of other substances, such as waxes and lipoproteins, giving the structures the properties of impermeability. Such chitinous structures have been extensively exploited by animals in the development of a number of different morphological systems or devices, assigned to a number of different functions. The biosynthesis of chitin–protein complexes has been subjected, during the course of animal evolution, to a number of "morphological radiations"[30,131].

Chitinous structures have been primarily exploited as protective envelopes, forming the theca and cyst walls of some Rhizopoda and Ciliata, the periderm or perisarc of Hydrozoa and Endoprocta, the tubes of Phoronida and Pogonophora, the shells of inarticulate Brachiopoda and of the molluscs. Chitinous structures are also used as envelopes of eggs or of latent forms of life such as cysts. By providing an adequate support for muscle insertions, chitinous structures contribute to the formation of locomotor appendages or chaetae and of prehensile and masticating organs such as jaws and radula. In Brachiopoda, a chitinous cuticle is used to insure the fixation of the organism to the substrate. Some buyoancy organs are also built up of chitin, as in Siphonophora and Cephalopoda. The chitinous peritrophic membrane of Arthropoda and annelids plays a protective role with respect to the intestinal mucosa and a role in the formation of the faeces.

The chitin–protein framework seems to provide a convenient support for calcification and silicification; as a result of this process of mineralisation of their chitinous structures, the organisms appear to realize an economy of protein material, as do many crustaceans, or an economy of both chitin and proteins, as in the case of molluscs. On the other hand, calcified exoskeleton also provides a rigid structure insuring the stability of colonies such as in Hydrozoa and Bryozoa.

At the top of the evolutionary lineage of protostomian invertebrates, the chitin–protein complexes in their most stable tanned form allow the formation of rigid planes at the expense of a minimal weight of material. Insects have developed extended portions of their teguments (paranota) which, owing to their rigidity and to their low weight, have successfully been transformed into functional wings. The wing hinges have the particular properties of a rubber-like cuticle, due to the linkage of chitin to resilin molecules. The realization of the flight system by insects is actually the consequence of the particular features of the different types of glycoprotein complexes that chitin can realize by linkage with arthropodin, sclerotin or resilin.

The realization of a complete chitinous cuticular envelope creates for the Arthropoda problems of growth and development, as well as a problem of epicuticle replacement after abrasion. The ability to synthesize chitinolytic and proteolytic enzymes kept at the level of the epidermis, and the hormonal control of a cyclic secretion of these enzymes are biochemical features which allowed the solution of these problems and explain the success encountered by such types of chitin-covered animals.

REFERENCES

1 A. HACHETT, *Phil. Trans. Roy. Soc. (London)*, (1799) 315.
2 A. ODIER, *Mém. Soc. Hist. Nat. (Paris)*, 1 (1823) 29.
3 R. H. HACKMAN, *Australian J. Biol. Sci.*, 13 (1960) 568.
4 D. HORTON AND M. L. WOLFROM, in M. FLORKIN AND E. H. STOTZ (Eds.), *Comprehensive Biochemistry*, Vol. 5, Elsevier, Amsterdam, 1963, p. 185.
5 A. B. FOSTER AND J. M. WEBBER, *Advan. Carbohydrate Chem.*, 15 (1960) 371.
6 P. W. KENT, in M. FLORKIN AND H. S. MASON (Eds.), *Comparative Biochemistry*, Vol. 7, Academic Press, New York, 1964, p. 93.
7 J. S. BRIMACOMBE AND J. M. WEBBER, *Mucopolysaccharides, (B.B.A. Library, Vol. 6)*, Elsevier, Amsterdam, 1964, p. 18.
8 H. HOMANN, *Z. Vergleich. Physiol.*, 31 (1949) 413.
9 R. O. HERZOG, *Angew. Chem.*, 39 (1926) 297.
10 C. J. B. THOR AND W. F. HENDERSON, *Am. Dyestuff Reptr.*, 29 (1940) 461.
11 D. F. WATERHOUSE, R. H. HACKMAN AND J. W. MCKELLAR, *J. Insect Physiol.*, 6 (1961) 96.
12 R. H. HACKMAN AND M. GOLDBERG, *Australian J. Biol. Sci.*, 18 (1965) 935.
13 C. H. GILES, A. S. A. HASSAN, M. LAIDLAW AND R. V. R. SUBRAMANIAN, *J. Soc. Dyers Colourists*, 74 (1958) 647.
14 K. M. RUDALL, *Advan. Insect Physiol.*, 1 (1963) 257.
15 R. O. HERZOG, *Naturwissenschaften*, 12 (1924) 955.
16 H. W. GONELL, *Z. Physiol. Chem.*, 152 (1926) 18.
17 K. H. MEYER AND H. MARK, *Chem. Ber.*, 61 (1928) 1932.
18 K. H. MEYER AND G. W. PANKOW, *Helv. Chim. Acta*, 18 (1935) 589.
19 D. CARLSTRÖM, *J. Biophys. Biochem. Cytol.*, 3 (1957) 669.
20 A. N. J. HEYN, *Koninkl. Ned. Akad. Wetenschap., Proc.*, 39 (1936) 132.
21 A. N. J. HEYN, *Protoplasma*, 25 (1936) 372.
22 G. FRAENKEL AND K. M. RUDALL, *Proc. Roy. Soc. (London)*, Ser. B, 134 (1947) 111.
23 W. LOTMAR AND L. E. R. PICKEN, *Experientia*, 6 (1950) 58.
24 K. M. RUDALL, *Symp. Soc. Exptl. Biol.*, 9 (1955) 49.
25 N. E. DWELTZ, *Biochim. Biophys. Acta*, 44 (1960) 416.
26 D. CARLSTRÖM, *Biochim. Biophys. Acta*, 59 (1962) 361.
27 G. N. RAMACHANDRAN AND C. RAMAKRISHNAN, *Biochim. Biophys. Acta*, 63 (1962) 307.
28 N. E. DWELTZ, *Biochim. Biophys. Acta*, 51 (1961) 283.
29 K. M. RUDALL, in *The Scientific Basis of Medicine, Annual Reviews*, Athlone, London, 1962, p. 203.
30 CH. JEUNIAUX, *Chitine et Chitinolyse*, Masson, Paris, 1963, 181 pp.
31 CH. JEUNIAUX, *Bull. Soc. Chim. Biol.*, 47 (1965) 2267.
32 A. G. RICHARDS, *The Integument of Arthropods*, Univ. Minnesota Press, Minneapolis, 1951.
33 M. V. TRACEY, in K. PAECH AND M. V. TRACEY (Eds.), *Modern Methods of Plant Analysis*, Vol. 2, Springer, Berlin, 1955, p. 264.
34 F. VON WETTSTEIN, *Handbuch der Systematischen Botanik*, 4th ed., Deuticke, Leipzig, 1933.
35 J. W. FOSTER, *The Chemical Activities of Fungi*, Academic Press, New York, 1949.
36 A. B. FOSTER AND M. STACEY, in W. RUHLAND (Ed.), *Encyclopaedia of Plant Physiology*, Part VI, Springer, Berlin, 1958, p. 518.
37 H. J. BLUMENTHAL AND S. ROSEMAN, *J. Bacteriol.*, 74 (1957) 222.
38 N. SHARON, in E. A. BALAZS AND R. W. JEANLOZ (Ed.), *The Amino Sugars*, Vol. 2a, Academic Press, New York, 1965, p. 1.

39 M. W. MILLER AND H. J. PHAFF, *J. Microbiol. Serol.*, 24 (1958) 225.
40 M. E. WURDACK, *Ohio J. Sci.*, 23 (1923) 181.
41 I. B. SACHS, *Trans. Am. Microscop. Soc.*, 75 (1956) 307.
42 M. SCHMIDT, *Arch. Mikrobiol.*, 7 (1936) 241.
43 R. H. HEDLEY, *Quart. J. Microscop. Sci.*, 101 (1960) 279.
44 M. WILFERT AND W. PETERS, *Z. Morphol. Oekol. Tiere*, 64 (1969) 77.
45 W. PETERS, *Zool. Anz.*, Suppl. 31 (1968) 681.
46 S. A. WAINWRIGHT, *Experientia*, 18 (1962) 18.
47 D. M. CHAPMAN, in W. J. REES (Ed.), *The Cnidaria and Their Evolution*, Symp. Zool. Soc., No. 16, Academic Press, London, 1966, p. 51.
48 D. M. CHAPMAN, *J. Marine Biol. Assoc. (U.K.)*, 48 (1968) 187.
49 H. DEPOORTERE AND N. MAGIS, *Ann. Soc. Roy. Zool. Belg.*, 97 (1967) 187.
50 J. L. CRITES, *Ohio J. Sci.*, 58 (1958) 343.
51 W. P. ROGERS, *Nature*, 181 (1958) 1410.
52 T. VON BRAND, *J. Parasitol.*, 26 (1940) 301.
53 D. B. CARLISLE, *Arkiv Zool.*, 2 (1958) 79.
54 D. BAY AND CH. JEUNIAUX, in preparation.
55 L. H. HYMAN, *Biol. Bull.*, 114 (1958) 106.
56 Y. SAUDRAY AND M. BOUFFANDEAU, *Bull. Inst. Oceanog.*, No. 1119 (1958).
57 L. E. R. PICKEN, *Biol. Rev.*, 15 (1940) 133.
58 M. DESIERE AND CH. JEUNIAUX, *Ann. Soc. Roy. Zool. Belg.*, 98 (1968) 43.
59 CH. JEUNIAUX AND C. TONNELIER, in preparation.
60 N. W. RUNHAM, *Ann. Histochem.*, 8 (1963) 433.
61 G. GOFFINET AND CH. JEUNIAUX, *Cahiers Biol. Marine*, 1970 (in preparation).
62 N. W. RUNHAM, *Quart. J. Microscop. Sci.*, 102 (1961) 371.
63 N. W. RUNHAM, *J. Histochem. Cytochem.*, 10 (1962) 504.
64 G. TOTH AND L. ZECHMEISTER, *Nature*, 144 (1939) 1049.
65 Y. HASHIMOTO AND T. HIBYA, *Bull. Japan. Soc. Sci. Fisheries*, 19 (1953) 5.
66 T. HIBYA, E. IWAI AND Y. HASHIMOTO, *Bull. Japan. Soc. Sci. Fisheries*, 19 (1953) 1.
67 G. GOFFINET AND CH. JEUNIAUX, *Comp. Biochem. Physiol.*, 29 (1969) 277.
68 H. STEGEMANN, *Z. Physiol. Chem.*, 331 (1963) 269.
69 C. BERKELEY, *Biol. Bull.*, 68 (1935) 107.
70 M. LAFON, *Ann. Sci. Nat., Botan. Zool.*, 11 (1943) 113.
71 M. LAFON, *Bull. Inst. Oceanog.*, 939 (1948) 28.
72 P. DRACH AND M. LAFON, *Arch. Zool. Exptl. Gen.*, 82 (1942) 100.
73 G. FRAENKEL AND K. M. RUDALL, *Proc. Roy. Soc. (London) Ser. B*, 129 (1940) 1.
74 T. WEIS-FOGH, *J. Exptl. Biol.*, 37 (1960) 889.
75 T. WEIS-FOGH, *J. Mol. Biol.*, 3 (1961) 520.
76 M. F. DAY, *Australian J. Sci. Res.*, (1949) 421.
77 R. DE METS AND CH. JEUNIAUX, *Arch. Intern. Physiol. Biochim.*, 70 (1962) 93.
78 J. BLACKWELL, K. D. PARKER AND K. M. RUDALL, *J. Marine Biol. Assoc. (U.K.)*, 45 (1965) 659.
79 P. C. BRUNET AND D. B. CARLISLE, *Nature*, 182 (1958) 1689.
80 M. F. FOUCART, S. BRICTEUX-GRÉGOIRE AND CH. JEUNIAUX, *Sarsia*, 20 (1965) 35.
81 W. PETERS, *Experientia*, 22 (1966) 820.
82 W. PETERS, *Z. Morphol. Oekol. Tiere*, 62 (1968) 9.
83 P. A. ROELOFSEN AND I. HOETTE, *J. Microbiol. Serol.*, 17 (1951) 297.
84 V. B. WIGGLESWORTH, *Quart. J. Microscop. Sci.*, 89 (1948) 197.
85 A. G. RICHARDS, in K. D. ROEDER (Ed.), *Insect Physiology*, Wiley, New York, 1953, p. 1.
86 M. LOCKE, *J. Biophys. Biochem. Cytol.*, 10 (1961) 589.

87 M. LOCKE, *Quart. J. Microscop. Sci.*, 98 (1957) 487.
88 W. PETERS, *Z. Zellforsch.*, 93 (1969) 336.
89 M. LOCKE, in M. ROCKSTEIN (Ed.), *The Physiology of Insecta*, Vol. 3, Academic Press, New York, 1964, p. 380.
90 K. BAILEY AND T. WEIS-FOCH, *Biochim. Biophys. Acta*, 48 (1961) 452.
91 S. R. A. MALEK, *Nature*, 180 (1957) 237.
92 S. R. A. MALEK, *Nature*, 185 (1960) 56.
93 B. M. JONES AND W. SINCLAIR, *Nature*, 181 (1958) 927.
94 H. F. LOWER, *Zool. Jahrb. Abt. Anat. Ontog. Tiere*, 76 (1957) 165.
95 H. F. LOWER, *Studium Generale (Berlin)*, 17 (1964) 275.
96 E. DUDICH, *Zoologica (Stuttgart)*, 30 (1931) 1.
97 D. F. TRAVIS, in *Calcification in Biological Systems*, A.A.A.S. Symposium, No. 64, Washington, 1960, p. 57.
98 D. F. TRAVIS, *Ann. N. Y. Acad. Sci.*, 109 (1963) 177.
99 D. F. TRAVIS AND U. FRIBERG, *J. Ultrastruct. Res.*, 9 (1963) 285.
100 D. F. TRAVIS, C. J. FRANÇOIS, L. C. BONAR AND M. J. GLIMCHER, *J. Ultrastruct. Res.* 18 (1967) 519.
101 R. DENNELL, *Proc. Roy. Soc. (London)*, Ser. B, 134 (1947) 485.
102 M. GABE AND M. PRENANT, *Bull. Lab. Maritime Dinard*, 37 (1952) 13.
103 E. I. JONES, R. A. MCCANCE AND L. R. B. SHACKLETON, *J. Exptl. Biol.*, 12 (1935) 59.
104 M. G. M. PRYOR, *Proc. Roy. Soc. (London)*, Ser. B, 128 (1940) 378.
105 M. G. M. PRYOR, *Proc. Roy. Soc. (London)*, Ser. B, 128 (1940) 393.
106 A. R. H. TRIM, *Nature*, 147 (1941) 115.
107 A. R. H. TRIM, *Biochem. J.*, 35 (1941) 1088.
108 L. H. JOHNSON, J. H. PEPPER, M. N. B. BANNING, E. HASTINGS AND R. S. CLARK, *Physiol. Zool.*, 25 (1952) 250.
109 R. H. HACKMAN AND M. GOLDBERG, *J. Insect Physiol.*, 2 (1958) 221.
110 B. W. DE HASS, L. H. JOHNSON, J. H. PEPPER, E. HASTINGS AND G. L. BAKER, *Physiol. Zool.*, 30 (1957) 121.
111 G. DUCHÂTEAU AND M. FLORKIN, *Physiol. Comp. Oecol.*, 3 (1954) 365.
112 H. H. MOOREFIELD, 1953, cited by H. H. HACKMAN, in M. ROCKSTEIN (Ed.), *The Physiology of Insecta*, Vol. 3, Academic Press, New York, 1964, p. 471.
113 A. G. RICHARDS, *Ann. Entomol. Soc. Am.*, 40 (1947) 227.
114 R. H. HACKMAN, *Australian J. Biol. Sci.*, 8 (1955) 83.
115 R. H. HACKMAN, *Australian J. Biol. Sci.*, 8 (1955) 530.
116 A. B. FOSTER AND R. H. HACKMAN, *Nature*, 180 (1957) 40.
117 M. M. ATTWOOD AND H. ZOLA, *Comp. Biochem. Physiol.*, 20 (1967) 993.
118 A. GOTTSCHALK, W. H. MURPHY AND E. R. B. GRAHAM, *Nature*, 194 (1962) 1051.
119 H. S. MASON, *Advan. Enzymol.*, 16 (1955) 105.
120 M. G. M. PRYOR, in M. FLORKIN AND H. S. MASON (Eds.), *Comparative Biochemistry*, Vol. 4, Academic Press, New York, 1962, p. 371.
121 R. Y. ZACHARUK, *Can. J. Zool.*, 40 (1962) 733.
122 G. KRISHNAN AND S. RAJULU, *Z. Naturforsch.*, 19 (1964) 640.
123 CH. JEUNIAUX, *Biochem. J.*, 66 (1959) 292.
124 CH. JEUNIAUX, *Arch. Intern. Physiol. Biochim.*, 67 (1959) 597.
125 A. C. NEVILLE, *Oikos*, 14 (1963) 1.
126 A. C. NEVILLE, *J. Insect Physiol.*, 9 (1963) 117.
127 A. C. NEVILLE, *J. Insect Physiol.*, 9 (1963) 265.
128 A. C. NEVILLE, *Quart. J. Microscop. Sci.*, 106 (1965) 269.
129 A. G. RICHARDS, *Ergeb. Biol.*, 20 (1958) 1.
130 A. G. RICHARDS AND R. L. PIPA, *Smithsonian Inst. Misc. Collections*, 137 (1958) 247.

131 W. V. COUDOLIS AND M. LOCKE, cited by M. LOCKE, in M. ROCKSTEIN (Ed.), *The Physiology of Insecta*, Vol. 3, Academic Press, New York, 1964, p. 379.
132 Y. BOULIGAND, *Compt. Rend.*, 261 (1965) 3665.
133 Y. BOULIGAND, *Compt. Rend.*, 261 (1965) 4864.
134 M. LOCKE, *Quart. J. Microscop. Sci.*, 101 (1960) 333.
135 E. A. ROBSON, *Quart. J. Microscop. Sci.*, 105 (1964) 281.
136 A. G. RICHARDS, *J. Morphol.*, 96 (1955) 537.
137 A. G. RICHARDS AND F. H. KORDA, *Biol. Bull.*, 94 (1948) 212.
138 W. HUBER AND C. HAASSER, *Nature*, 165 (1950) 397.
139 E. H. MERCER AND M. F. DAY, *Biol. Bull.*, 103 (1952) 394.
140 W. PETERS, *Z. Morphol. Oekol. Tiere*, 64 (1969) 21.
141 R. DE METS, *Nature*, 196 (1962) 77.
142 K. M. RUDALL, *Aspects of Insect Biochemistry*, Academic Press, New York, 1966, p. 83.
143 Y. BOULIGAND, *Z. Zellforsch.*, 79 (1967) 332.
144 CH. GRÉGOIRE, G. DUCHATEAU AND M. FLORKIN, *Ann. Inst. Océanog.*, 31 (1955) 1.
145 CH. GRÉGOIRE, *J. Biophys. Biochem. Cytol.*, 3 (1957) 797.
146 CH. GRÉGOIRE, *Bull. Inst. Roy. Sci. Nat. Belg.*, 37 (1961) No. 3, 1.
147 CH. GRÉGOIRE, *Bull. Inst. Roy. Sci. Nat. Belg.*, 38 (1962) No. 49, 1.
148 G. GOFFINET AND CH. JEUNIAUX, *Comp. Biochem. Physiol.*, 29 (1969) 277.
149 G. GOFFINET, *Comp. Biochem. Physiol.*, 29 (1969) 835.
150 M. FLORKIN, *A Molecular Approach to Phylogeny*, Elsevier, Amsterdam, 1966, 176 pp.
151 M. LOCKE, in J. W. L. BEAMENT AND J. E. TREHERNE (Eds.), *Insects and Physiology*, Edinburgh, 1967.
152 J. V. PASSONNEAU AND C. M. WILLIAMS, *J. Exptl. Biol.*, 30 (1953) 545.
153 P. DRACH, *Ann. Inst. Océanog.*, 19 (1939) 106.
154 C. D. COTTRELL, *J. Exptl. Biol.*, 39 (1962) 431.
155 C. D. COTTRELL, *J. Exptl. Biol.*, 39 (1962) 449.
156 C. D. COTTRELL, *Advan. Insect Physiol.*, 2 (1964) 175.
157 L. GLASER AND O. H. BROWN, *J. Biol. Chem.*, 228 (1957) 729.
158 E. G. JAWORSKI, L. WANG AND G. MARCO, *Nature*, 198 (1963) 790.
159 F. G. CAREY, *Comp. Biochem. Physiol.*, 16 (1965) 155.
160 D. J. CANDY AND B. A. KILBY, *J. Exptl. Biol.*, 39 (1962) 129.
161 C. A. PORTER AND E. G. JAWORSKI, *J. Insect Physiol.*, 11 (1965) 1151.
162 H. F. LOWER, *Am. Midland Naturalist*, 61 (1959) 390.
163 D. F. TRAVIS, *Biol. Bull.*, 109 (1955) 484.
164 CH. JEUNIAUX, *Arch. Intern. Physiol. Biochim.*, 58 (1955) 114.
165 CH. JEUNIAUX, AND M. AMANIEU, *Experientia*, 9 (1955) 195.
166 L. RENAUD, *Ann. Inst. Océanog.*, 24 (1949) 260.
167 M. R. LUNT AND P. W. KENT, *Biochim. Biophys. Acta*, 44 (1960) 371.
168 CH. JEUNIAUX, *Arch. Intern. Physiol. Biochim.*, 67 (1959) 516.
169 CH. JEUNIAUX, *Arch. Intern. Physiol. Biochim.*, 68 (1960) 684.
170 CH. JEUNIAUX, *Mém. Soc. Roy. Entomol. Belg.*, 27 (1955) 312.
171 E. ABDERHALDEN AND K. HEYNS, *Biochem. Z.*, 259 (1933) 320.
172 D. B. CARLISLE, *Biochem. J.*, 90 (1964) 1c.

Chapter X

Resilin

S. O. ANDERSEN

Department of Zoology, University of Cambridge (Great Britain)

1. General description and definition

Resilin is a structural protein which shows long-range rubber-like elasticity and is found in insects. It was first described by Weis-Fogh in 1960[1], and it has since been analysed in detail with respect to mechanical and chemical properties. The name resilin is derived from the Latin *resilire*, to spring back[2]. A working definition of resilin was suggested in 1964[3], and only minor changes need now be made in this definition due to the identification of the cross-links present in resilin.

Pure resilin as found in dragonfly and locust is a colourless, transparent material which shows no structure in light microscopy or electron microscopy[4] and is completely insoluble in all nondestructive solvents. It is optically isotropic when unstrained in the swollen state, and it shows rubber-like long-range elasticity without any tendency to lasting deformations even after prolonged stretching. When strained it shows birefringence which is positive in the direction of stretch[1].

Resilin stains blue with methylene blue and toluidine blue without any indication of metachromasia, and it stains red with the standard histological staining reactions of Masson and Mallory[1,3].

Resilin shows a strong blue fluorescence when exposed to ultraviolet light. The wavelength of maximum emission is about 410 μ in both acid and alkaline media, and the wavelength of maximum excitation is about 280 mμ in the acid range and about 320 mμ in the alkaline range. The intensity of the fluorescence increases considerably in going from acid to alkaline medium, and the change takes place near neutrality[5,6].

[633]

Resilin is a simple protein in so far as it only gives rise to amino acids on hydrolysis. Two of these are unusual, being a dimer and a trimer of tyrosine[6,7], and they are responsible for the characteristic fluorescence of resilin. Several of the usual amino acids have not been found in resilin, including tryptophan, cystine, and methionine; hydroxyproline and hydroxylysine are also absent[2].

It must be emphasized that this description of resilin only applies to samples of *pure* resilin. In samples where chitin is present the mechanical properties of the structure will be modified although the chemical properties will be unchanged. If chitin is present as lamellae the structure will be prevented from swelling isotropically and the resilin will become birefringent due to internal strains. The sample cannot be stretched easily in directions parallel to the plane of the lamellae, whereas they will give only little resistance to compression or bending.

If other structural proteins are present this will not only influence the mechanical properties but the chemical properties as well. Staining methods cannot therefore be relied upon in such cases, and the fluorescence cannot be trusted as an indication of the presence of resilin as the other compounds present may either quench the fluorescence or show autofluorescence in the visible range. Isolation of the two specific amino acids which give rise to the fluorescence of resilin will in such cases be a much better indication for the presence of resilin. Preferably one should be able either to remove the interfering material and show that the remaining material is much more resilin-like than the starting material, or one should be able to remove the resilin specifically. This, however, is difficult, as resilin is completely insoluble as long as it has not been broken down to smaller peptides, so it can only be removed by rather drastic extraction procedures. Mild acid hydrolysis appears to hydrolyze resilin faster than the other cuticular proteins in locusts presumably due to the high content of aspartic acid in resilin, as aspartyl peptide bonds are known to be specially labile in dilute acid[8]. Hydrolysis with 0.1 M hydrochloric acid at 100° for 6 h has been used routinely to remove resilin from chitin which appears to be unaffected by this treatment[2,5].

2. Occurrence of resilin

Resilin was first found in two types of ligaments involved in the movements of the wings during flight of locusts (*Schistocerca gregaria*), and in an elastic tendon in dragonflies (*Aeshna*)[1]. It is these three test pieces which have solely been used in the investigations of the properties of resilin and which have formed the basis for the definition of resilin, although resilin is found in many other structures as well.

The elastic tendon from dragonflies is the most regularly shaped of all the samples forming a hollow tube-like swelling on the tendon which connects the pleuro-subalar muscle to the ventral wall. The swelling is about 0.7 mm long and 0.15 mm wide in big *Aeshna* species and contains 5–7 μg of resilin. Apart from a thin epicuticle covering the air-filled central canal and an equally thin cortical membrane the whole swollen part of the tendon consists of resilin which gives all the typical resilin reactions. Chitin is only present in the tough parts of the tendon which anchors the swollen part both to the muscle and to the ventral cuticle. Chitin may also be present in the thin epicuticle covering the central canal as a faint positive reaction was obtained with the chitosan test[1]. Epicuticle in insects is, however, supposed to contain no chitin[9].

The simplest of the major resilin-containing structures in locusts is the so-called prealar arm which connects the anterior end of the movable plate between the forewings (the mesotergum) to the first basalar sclerite of the thoracic wall. The prealar arm is formed like a small cone with a V-shaped groove on one side. The base of the cone goes abruptly over into the hard sclerotized cuticle of the mesotergum whereas the tip of the cone is connected to the sclerotized cuticle of the thorax wall through a tough flexible ligament[1]. The cone is about 0.5 mm long and 0.2–0.3 mm wide, and it contains about 0.05 mg resilin and about 0.015 mg chitin[10]. The chitin is described as occurring in thin concentric lamellae, less than 0.2 μ thick[4]. The chitin lamellae are arranged parallel to the inner, concave side of the cone, so that the cone is easily bent but resists stretching[1,11].

However, Weis-Fogh[12] now favours the idea that chitin is organized somewhat similar to the helicoidal model originally suggested by Bouligand[13] for crustacean cuticle and which later also has been applied to solid insect cuticle[14]. According to this model the chitin molecules are organized in sub-microscopical layers, the molecules being parallel in each layer and a small progressive change in angle of orientation occurring from one layer to the next. When oblique sections of such a structure are viewed in a microscope they will appear lamellate.

In locust the largest structure containing resilin is the main hinge ligament from the forewings. It is a thick irregularly shaped piece of hyaline cuticle located between the hard sclerotized mesopleural wing process and the similarly sclerotized second axillary wing sclerite. The transition between the rubbery ligament and the sclerotized cuticles is very abrupt whereas there is a gradual transition from the sides of the ligament into tough flexible arthrodial membrane. The resilin content of the ligament is about 0.10 mg and the chitin content is about 0.02 mg[10]. The main part of the ligament appears like the prealar arm to be composed of alternating thin chitin lamellae and thicker layers of resilin. However, a part of the ligament located on the epidermal cell side is deposited as a pad of pure resilin without chitin being present[1,10].

Resilin is also present in locust in some minor wing ligaments, where it occurs together with chitin but separated clearly from the sclerotized cuticle, whereas in the clypeo-labral spring and the abdominal tergites of locust there is no clear separation between resilin and sclerotized cuticle. The clypeo-labral spring is a pair of rib-shaped swellings on the oral aspect of clypeus and labrum. These plates cover the mouth-parts anteriorly, and the spring serves to keep the labrum against the mandibles[3]. The spring shows long-range elasticity, it shows the staining reactions and fluorescence characteristic of resilin, and di- and trityrosine have been isolated from hydrolysates of the spring. When sections are viewed in the fluorescence microscope it is seen that the characteristic blue resilin fluorescence gradually disappears when going from the centre of the spring towards the surrounding hard cuticle. The term "transitional cuticle" has been suggested for this intermediate type of cuticle, which contains resilin but where other cuticular proteins are also present[3].

A similar situation is found in the cuticle of the abdominal tergites. Here the fluorescence is found in the midline of the tergites and in a region along the hind margin[3]. These regions also give the typical resilin-staining reactions and di- and trityrosine have been isolated from hydrolysates of the regions, so there is little doubt that they contain resilin. However, there is no sharp demarcation between the fluorescent regions and the surrounding tergite cuticle.

An interesting example of a resilin-containing structure is the soft cuticle covering the tarsal and aroliar pads in locust. The detailed structure of the pads has not yet been investigated, and the identification of resilin is based on the autofluorescence of the cuticle, its elasticity, and the isolation of di- and trityrosine from hydrolysates of the pads[6]. It is not known whether the cuticle of the pads is of the transitional type or whether resilin is the only protein present.

In the foot pads of the fleshfly *Sarcophaga bullata* Witten[15] has recently described a cuticular layer, classified as endocuticle, which she reports to be remarkably similar to resilin, but all her attempts to determine the presence of resilin have proved equivocal.

Resilin has also been reported to be present in elastic structures involved in the jumping of insects. Bennet-Clark and Lucey[16] have analysed the jumping mechanism of fleas, and they calculate that the energy requirement for a jump of 3.5 cm height of the rabbit flea, *Spilopsyllus cuniculi*, is about 2.25 erg. This energy must be delivered in 0.75–1.0 msec over a distance of 0.37–0.5 mm, which is not compatible with direct muscle action, whereas it can be performed by means of an elastic spring in which sufficient energy has been stored. The authors describe the presence of a small resilin pad located between the notum and pleuron. The pad has an elliptical shape of nearly constant thickness, it is 85–100 μ high, 65–75 μ long and has an average volume of $1.4 \cdot 10^{-4}$ mm^3. Its energy-storing potential is calculated to be about 2.1 erg per pad, 4.2 erg per flea, which is somewhat more than the amount of energy needed for an average jump. According to Neville and Rothschild[17] the pad gives the typical staining reactions for resilin and is autofluorescent with the same blue colour as locust resilin. Chitin lamellae are only present in the part deposited first and pore canals are absent. The pad was found to be homologous to the wing-hinge ligament in locust and other insects[17].

Resilin also appears to be involved in the jumping of click beetles: Sannasi[18] has found that the endocuticle of the mesosternal plate of *Melanotus communis* stains like resilin with toluidine blue, swells without going into solution in several organic solvents, and shows a strong blue fluorescence which becomes stronger in alkaline media, and that both di- and trityrosine are present in hydrolysates of the cuticle according to paper chromatography.

In beetles resilin has also been found present in the so-called abdominal springs[3], which are rib-shaped swellings of the endocuticle running from the lateral side of the tergites across the soft arthrodial membrane and ending on the sternites. The function of these elastic ribs is to keep the tergites and sternites in the inspiratory position when the expiratory muscles are relaxed[3].

Clements and Potter[19] have reported the presence of resilin in the cuticle lining the spermathecal duct in the mosquito *Aedes aegypti*. Part of the cuticle appeared homogeneous in the electron microscope, and the duct is elastic, shows a blue autofluorescence and it becomes birefringent on stretching. Gupta and Smith[20] in a study of the structure of the spermatheca in the cockroach, *Periplaneta americana*, suggest that this animal also may have resilin in the wall of the spermathecal duct.

Resilin is involved in the opening mechanism for spiracle 2 in locust according to Miller[21]. The spiracle possesses a single closer muscle in contrast to spiracles 1 and 3 which have both closer and opener muscles. The elasticity of the patch of resilin is partly responsible for the opening of the spiracle during flight.

All the instances of the occurrence of resilin in insects discussed so far concern structures where long-range elasticity and nonpermanent deformation are important for the proper function of the structures. The presence of resilin in a structure where elasticity would not be expected to be of importance has only been reported in a single case. Recently Sannasi[22] described the presence of what appears to be resilin in the cuticle covering the compound eyes in the firefly *Photinus*. The identification is again based on the staining properties, the autofluorescence, and the presence of di- and trityrosine. Andersen and Weis-Fogh[3] classified the endocuticle from locust eyes as transitional cuticle, but I have not been able to obtain any di- and trityrosine from hydrolysates of this cuticle.

Only a single example is known of resilin being present in other animals than insects. It has been found in the elastic leg-hinge of the crayfish, *Astacus fluviatilis*[3], where it functions as a leg extensor since the hinge is equipped with a flexor muscle but has no extensor muscle.

Weis-Fogh has looked for resilin in several species of spiders without finding it, whereas to my knowledge no one has yet investigated other groups of arthropods, so we do not know how widespread the occurrence of resilin is.

Although resilin itself has not been found in vertebrates, the occurrence of one of the specific cross-links from resilin, dityrosine, has been reported present in preparations of vertebrate elastin[23]. From hydrolysates of elastin LaBella and collaborators isolated a compound, called "U", which from labelling experiments could be shown to be derived from tyrosine. It was identified as dityrosine on the basis of its fluorescence maximum and its chromatographic behaviour on ion-exchange resins relative to authentic dityrosine. It has later been reported that the dityrosine does not originate from elastin itself but from an alkali-soluble protein present in elastic tissue[24].

Apart from tyrosine several other phenolic amino acids have been found present in hydrolysates of the structural proteins (elastoidin) present in the fins of elasmobranchs[25]. These amino acids have not yet been identified, two of them show R_F values similar to di- and trityrosine, but they are non-fluorescent[26]. Whether these amino acids have anything to do with crosslinking of the proteins in elastoidin is not known. Ignac and Joseph[27] have also reported the absence of dityrosine from elastoidin, although they found that significant amounts of dityrosine are formed when elastoidin is treated with hydrogen peroxide.

Although dityrosine may thus be present in some component of elastic tissue in vertebrates there is no evidence indicating that this component is resilin. The presence of dityrosine cannot be sufficient to classify a protein as resilin, most of the other properties must also correspond to those described for the resilin-containing structures in locust and dragonfly.

In summary of this list of the presence and absence of resilin and resilin-like material in different animals it can be concluded that resilin as such has only been found in special cuticular structures in arthropods, and apparently it is not present in all arthropods. It is mainly found in regions where elasticity is necessary for the function of the structure, and it can occur both as pads of pure resilin, as lamellate cuticle with layers of resilin alternating with lamellae of chitin, or it can be present together with both chitin and proteins of a nonresilin-like nature in transitional cuticle.

3. Physical and chemical properties of resilin

As mentioned above, resilin is completely insoluble in all solvents which do not break the peptide chains although several solvents are able to penetrate into the resilin matrix and make the material swell[1,3]. The best swelling agents are those which are also good solvents for proteins in general, such as formamide, phenol, formic and acetic acid and ethanolamine. In all these solvents pieces of pure resilin swell more than they do in distilled water. In ethylene glycol, glycerol, dimethylformamide, and dimethylsulphoxide the swelling is similar to that in water, and resilin does not swell in such solvents as methanol, ethanol, acetone, diethylether and dioxan. When transferred back from any of these solvents to pure water resilin returns to its initial degree of swelling, and no permanent changes have occurred. Autoclaving pieces of resilin at 120° in water also leaves them completely unaffected, indicating that resilin cannot be heat denatured.

The swelling of resilin in water at different pH values shows that there is a pronounced effect[1,3] of pH. A minimum of swelling is found around pH 4, the isoelectric point of the protein, and the swelling increases much more on the alkaline side of the isoelectric point than on the acid side. No irreversible change takes place in the material in the pH range from 1 to 12, and it retains its rubber-like properties at all pH values[11,28].

Although all samples of resilin show long-range elasticity it is only in the case of the dragonfly tendon that it has been possible to analyse fully the elasticity by means of mechanical and optical methods[28,29]. Some measurements have also been made on the prealar arm from locust[11], but due to its more irregular shape and the presence of chitin lamellae it has not been possible to obtain as detailed an analysis as for the dragonfly tendon.

The measurements on the tendon were performed by having it suspended under the microscope in such a way that known external forces could easily be applied to the sample and the resulting changes in length measured. The composition and the temperature of the bathing medium could be changed at any time. Since only swollen resilin shows long-range elasticity, and the sample therefore is a thermodynamically open system, it is necessary to find experimental conditions where there is little or no change in swelling with changes in temperature. A transfer of solvent between medium and sample would make the calculation of the thermodynamic parameters of the system difficult. The tendon showed volume changes of less than 3% over a temperature range of 62° when placed in dilute phosphate buffer at pH 6.7, and in this medium the isometric force of the tendon increases linearly with absolute temperature at all degrees of stretching. This is the behaviour to be expected from an ideal rubber where the elastic force is solely due to thermal agitation of randomly flexible chains.

For different degrees of stretching Weis-Fogh calculated how much of the force is due to an increase in the internal energy of the protein and how much is due to a decrease in configurational entropy of the chains. The result showed that for extensions ranging from 1.1 to 2.6 the increase in internal energy is small, and that the entropy term can account for nearly the total force used for extension[29].

The results are in complete agreement with the properties of an ideal rubber, and Weis-Fogh[28] therefore compared the experimental force–extension curves with the theoretical behaviour according to a kinetic theory for a short-chain rubber network[30]. One of the main assumptions of the theory is that a rubber consists of an isotropic three-dimensional network of flexible chain molecules kept together by means of a few stable cross-links. The chain segments between the cross-links are thermally agitated and tend to take up the statistically most probable configuration. There is a simple relationship between the elastic modulus, which can be obtained from the experimental curves, and the average molecular weight of the chain segments between two cross-links. The latter value can therefore be calculated if the properties of the material are in accordance with the theory for rubber-like materials.

For dragonfly tendon this was found to be the case, and the elastic modulus was found to be about 6–7 kg/cm^2, corresponding to an average molecular weight of about 5000 for the chain segments between cross-links. This means that the peptide chains contain about 60 amino acid residues between two cross-links. For the prealar arm from locusts an elastic modulus of about 9 kg/cm^2 has been obtained[11], which corresponds to a molecular weight of 3400 for the chain segments or an average of 40 amino acid residues being present between two cross-links.

From the measurements on tendon[28] it was also possible to estimate the degree of flexibility of the chain segments and this was found to correspond to the presence of about 15 random links with full freedom of rotation between two junction points. Such a high degree of rotational freedom can only mean that there can be little constraint imposed upon the chains and their mutual interaction in the swollen state must be low.

The conclusions drawn from the mechanical measurements were confirmed by measuring the changes in birefringence on stretching the tendon; a close agreement was obtained between theoretically predicted values and those measured[28].

The dynamic properties of the prealar arm from locust have been investigated by exposing it to a sinusoidal strain with varying frequencies[11], and the elastic loss factor was determined. The loss factor is the fraction of the energy used for deforming the system which is lost as heat during the deformation and which cannot therefore be released as mechanical work. When the deformations were kept within the biological range the loss factor was only about 3%, which is lower than the values found for most natural and synthetic rubbers.

In all its mechanical and optical properties resilin thus behaves as a nearly ideal rubber, and Weis-Fogh[28] concluded that the molecular structure of resilin is that of randomly coiled protein chains having no stable secondary structure and being kept in continuous movement due to the thermal agitation of the chains. The chains will have to be firmly linked to each other by means of stable cross-links between them.

This picture of the resilin structure was confirmed by X-ray-diffraction measurements and electron microscopy[4]. In X-ray-diffraction measurements dry, unstretched tendon did not show any other reflections than a diffuse ring with a spacing of about 4 Å, indicating that resilin is a truly amorphous material. Tendon which had been stretched to about three times the unstrained length and then very slowly dried to allow the maximum amount of crystallization to take place showed the same diffuse ring as unstretched tendon and in addition also a single equatorial reflection corresponding to 4.5 Å. The appearance of this reflection is completely reversible. It disappears if the stretched tendon is rewetted, allowed to relax, and then redried in the relaxed position. No reflections could be found in low-angle diagrams for either stretched or unstretched samples.

Electron microscopy of osmium-fixed and methacrylate-embedded samples of resilin from both dragonfly and locust also showed that resilin is an amorphous material[4]. No trace of any fine structure could be found, the resilin looks as unstructured as the methacrylate-embedding medium. It appears that fixatives fail to stabilize resilin, and that the embedding medium does not penetrate into the resilin network, so that resilin in the sections used for electron microscopy becomes rubbery as soon as it comes into contact with polar solvents.

4. Amino acid composition of resilin

The amino acid composition of various types of locust rubber-like cuticle and of resilin from different species has been determined in order to see if there are any similarities in overall composition. In 1962 Bailey and Weis-Fogh[2] published some amino acid analyses for resilin which showed that this protein has certain characteristic features which set it apart from other structural proteins. Their results are given in Table I which also includes some new unpublished results, both for locust resilin and for resilin from other insects.

It can be seen that locust wing-hinges and prealar arms have similar amino acid compositions, and that there are no discrepancies between the old and new determinations. Resilin resembles many other structural proteins, such as collagen, elastin and silk fibroin in being very rich in

TABLE I

AMINO ACID COMPOSITION OF VARIOUS RESILINS

(Residues of amino acids per 100 residues)

	Schistocerca gregaria				Aeshna juncea Dragonfly tendon	Oryctes rhinoceros Abdominal spring
	Wing hinges		Prealar arms		Clypeo-labral spring	
	A^a	B	A^a	B			
Aspartic acid	9.9	10.2	10.8	11.3	7.5	9.4	22.6
Threonine	3.1	2.9	3.0	3.1	3.4	2.0	4.1
Serine	7.9	7.7	8.1	7.5	8.9	12.8	4.9
Glutamic acid	4.6	4.3	4.8	3.8	6.4	4.2	5.9
Proline	7.6	7.5	7.9	7.7	7.2	7.5	8.9
Glycine	39.1	41.1	36.7	39.7	31.4	42.2	33.5
Alanine	11.2	11.2	10.6	11.2	13.5	7.0	4.6
Valine	2.6	2.3	3.2	2.5	3.1	1.2	2.5
Cystine	—	—	—	—	—	—	—
Methionine	—	—	—	—	—	—	—
Isoleucine	1.7	1.3	1.7	1.4	1.9	0.9	2.6
Leucine	2.3	2.2	2.4	2.8	5.2	3.0	3.9
Tyrosine	2.5	2.1	3.1	1.2	1.9	1.2	0.5
Phenylalanine	2.6	2.5	2.4	2.4	2.8	2.1	2.2
Lysine	0.5	0.4 ⎫		⎧ 0.5	0.9	0.8	0.5
Histidine	0.9	0.8 ⎬	5.3	⎨ 1.1	0.9	1.2	0.5
Arginine	3.5	3.6 ⎭		⎩ 3.8	5.1	4.6	2.9
Sum of polar residues	32.9	32.0	35.1	32.3	35.0	36.2	41.9

a From Bailey and Weis-Fogh[2].

glycine residues, but it contains no hydroxyproline, hydroxylysine, trypto-phan, or any sulphur-containing amino acids. Resilin is relatively rich in proline (about 8%), and the amount of hydrophilic amino acid residues is high (about 33% of the residues).

The locust clypeo-labral spring, which is of the transitional type of cuticle with no sharp demarcation between the resilin-containing cuticle and the neighbouring hard cuticle, has an amino acid composition very similar to that of the ligaments, the main difference being that the glycine content is a little lower in the spring than in the ligaments. This might be due to the presence of hard-cuticle proteins since hard locust cuticle is low in glycine residues (8–12%, Table III).

The differences between resilin from various insect species are also small as can be seen by comparing the figures from locust ligaments and dragonfly tendon. The serine content is somewhat higher and alanine and valine a little lower in tendon than in ligaments, but there are no outstanding differences. The protein from the abdominal springs in Rhinoceros beetles, which from other criteria has been classified as resilin, also has a composition resembling locust resilin apart from the content of aspartic acid which is very high, 23% in the lateral spring as compared to 10% in locust resilin.

Although the amino acid composition of resilin-containing cuticle has only been obtained from three different species, it appears that the variation in composition of resilin between species is quite small. As more species are investigated a larger variation in amino acid composition will presumably be found, but it can be assumed that all resilins will be rich in amino acids with short side-chains and in hydrophilic amino acids and probably rich in proline[3]. Within these restrictions a wide range of variations should be expected.

In the discussion of amino acid composition of resilin no mention has been made of the unusual amino acids which are present. This is due to the fact that they have to be determined separately, as they are not easily determined during a normal amino acid analysis. Due to their importance for the structure of resilin they will be discussed separately in the next section.

5. Cross-links in resilin

The mechanical properties of resilin indicated that some sort of stable cross-links are present connecting the protein chains in a large three-dimensional network. As the properties of resilin are not changed by treatments which either reduce or oxidize disulphide bonds, it was concluded that the cross-links could not be cystine[1].

After a prolonged search for possible cross-links two unusual amino acids were isolated from hydrolysates of resilin[5] and were later identified as di- and trityrosine[6,7].

Dityrosine   Trityrosine

According to their structure they can function as cross-links in proteins since they can be built into two or three different chains by means of peptide linkages, and according to all available evidence this is their role in resilin.

A short description of the methods found suitable for the isolation of these amino acids will be given here. A more detailed account can be found in the earlier papers[6,7].

Resins made from sulphonated polystyrenes which are often used for the fractionation of amino acids are not suited for the separation and isolation of di- and trityrosine, since these amino acids tend to become adsorbed to the matrix of the resins, and they are therefore eluted late and in very broad peaks. A much better separation can be obtained on columns filled with cellulose phosphate in equilibrium with dilute acetic acid, or by means of DEAE-cellulose in equilibrium with dilute sodium bicarbonate[6]. In both cases the elution is performed by means of a sodium chloride gradient[6]. Di- and trityrosine can be located in the eluate by means of their fluorescence and they are well separated from each other and all other amino acids. They can thereafter be separated from the accompanying salts by passage through a column of BioGel P-2, elution being performed with dilute acetic acid. Another method for desalting solutions of di- and trityrosine is to adsorb them to a small column of cellulose phosphate in ammonium form, wash with distilled water and elute with dilute ammonia[6].

The compounds isolated in this way were shown to be homogeneous according to paper chromatography in several solvents, and it was demonstrated that one of them (dityrosine) has two α-amino groups and two carboxylic groups and that the other (trityrosine) contains three of each of these groups[5].

The identification of the two compounds as di- and trityrosine is mainly based on comparison with the synthetic compounds prepared according to the method of Gross and Sizer[31] using treatment of tyrosine with hydrogen peroxide and peroxidase from horseradish. Di- and trityrosine were purified from the reaction mixture by using the same fractionation scheme as for the compounds isolated from resilin. The natural and synthetic compounds were not only identical with respect to ultraviolet and fluorescence spectra but also with respect to chromatography on columns and paper in different solvents. The synthetic and natural dityrosine have also been compared with respect to the mass spectra of the methylated and trifluoroacetylated derivatives, and no differences were found in the spectra[32].

Apart from di- and trityrosine a very small amount of a third compound has also been obtained from resilin hydrolysates; this compound has tentatively been identified as tetratyrosine[6].

Di- and trityrosine have characteristic ultraviolet spectra (Table II) identical to those found for synthetic di- and trityrosine. The change from the acid to the alkaline spectrum occurs at pH 7.2 and 6.3, respectively, for the two compounds[5].

TABLE II

WAVELENGTH OF MAXIMAL ABSORBANCE OF DI- AND TRITYROSINE ISOLATED FROM LOCUST RESILIN

	Dityrosine (mμ)	Trityrosine (mμ)
pH 3	283.5	286
pH 10	316	322
pH 13	316	315
Borate, pH 8	252 and 292	292

The fluorescence spectra are identical for the natural and synthetic compounds[6]. The excitation spectra corresponded closely to the ultraviolet spectra at all pH values, and the emission spectra had their maximum located between 410 mμ and 420 mμ for all the compounds and irrespective of the pH of the solutions (between pH 1 and 13). Lehrer and Fasman[33] give emission spectra for dityrosine at 410 mμ both at pH 3.8 and 10.7. Andersen[5] found that the fluorescent light emitted from whole pieces of resilin at slightly alkaline reaction had maximum at 420 mμ. Lehrer[34] has since been able to obtain emission spectra from a single prealar arm suspended on a thin stainless-steel wire. He found identical emission curves with maximum at 410 mμ in water, 0.01 N HCl and 0.01 N NaOH. The excitation spectra of single pieces of resilin showed maximum in acid medium at 290 mμ and in alkaline medium at 320 mμ, and the change from the acid form of the spectrum to the alkaline form took place at pH 6.4.

The fluorescent behaviour of the whole untreated ligament is thus very similar to that of the two isolated tyrosine derivatives. Enzymatic digests of resilin show the same fluorescent behaviour as the isolated compounds[6], indicating that the compounds are not artefacts formed during acid hydrolysis, but that they are present as such in the native protein. In the enzymatic digests the compounds are present in peptides, and although a single peptide containing one of the compounds has not yet been obtained in pure form making a sequence analysis possible, all evidence indicates that both amino groups in dityrosine and all three amino groups in trityrosine are involved in peptide linkages[6]. Dityrosine will thus link two peptides together and trityrosine will link three together. The only point which remains to be established is whether these two (three) peptides originate from different chains or whether they come from different parts of the same chain. In the latter case the result would be loops on the chains and not a crosslinked network.

The possibility that a few loops are present cannot be disregarded, although it is improbable that a main part of the di- and trityrosine residues present takes part in the formation of loops. The mechanical analysis indicated that the average molecular weight of the peptide-chain segments between two cross-links is close to 3400 in locust resilin (p. 641). The chemical determination of the amounts of di- and trityrosine present in locust resilin gave an average of 8.9 and 4.5 residues per 10^5 g resilin, respectively[6], corresponding to 0.77 and 0.39 residues per 100 total residues. From this amount of crosslinking amino acids the average distance along the peptide chain between two cross-links can be calculated. It is found to correspond[6] to a molecular weight of 3200. The close agreement between the predicted degree of crosslinking and that actually found indicates not only that di- and trityrosine are the cross-links in resilin but also that they are the only cross-links present.

6. Deposition of resilin

It was shown by Neville[10] in 1963 that the formation of resilin-containing structures in locusts is a relatively slow process starting a few days before the final moult and still continuing about three weeks later although at a much slower rate. During this period both resilin and chitin are being secreted from the epidermal cells and deposited extracellularly. The formation of di- and trityrosine runs parallel with the deposition of resilin as the total amount of these two amino acids is proportional to the amount of resilin during the whole period of deposition, indicating that resilin is being crosslinked immediately after secretion from the cells. This agrees with the finding that resilin is completely insoluble at all stages of the formation of the structures, in contrast to what is found for the solid cuticle (p. 653).

When injected into adult locusts during the weeks of maturation labelled tyrosine becomes incorporated into resilin, and the radioactivity is recovered not only as tyrosine but also as di- and trityrosine[35], demonstrating that these amino acids are formed from tyrosine. Autoradiography of sections of ligaments from locust given one or more injections of tritium-labelled tyrosine[36,37] shows both that the incorporation of activity into resilin is a fast process and that the turnover of tyrosine in the animal is rapid. Activity is found incorporated into the newly synthesized resilin 20–30 min after the injection, and two injections with a time interval of 6 h give rise to two narrow, clearly separated bands of activity.

Gupta and Weis-Fogh[38] have used isolated prealar arms, covered with resilin- and chitin-producing epidermal cells, to achieve a true pulse labelling. Prealar arms were incubated for 3 min in suitable media containing tritiated tyrosine followed by a chase with unlabelled tyrosine for varying periods. Incorporated activity was analysed by high-resolution autoradiography and electron microscopy. They found that after only 5 min of incorporation the radioactivity is almost entirely concentrated in the Golgi regions of the epidermal cells, from where it is apparently transported to complex vesicles near the apical border of the cells. The activity begins to appear in extracellular resilin after about 10 min of total incubation.

Radioactivity which has become incorporated into the resilin apparently stays there for the rest of the animal's life as the bands obtained by autoradiography are as strong in animals killed several weeks after the injection as in those killed after a day or two, indicating that there is little or no turnover of resilin in adult animals after it has been deposited.

Neville[39,40] has made some further observations which support the idea that resilin is being deposited continuously and that it does not become modified after its deposition. He noticed that alternating strongly and weakly fluorescent bands can be seen in sections of the prealar arm and the wing-hinge by means of fluorescence microscopy. The number of band pairs present in young animals depends upon their age, and can be explained by the assumption that one band pair is deposited during each day–night period. It was demonstrated that the first three band pairs correspond to the resilin deposited during the last 3 days before the moult, the pre-moult resilin, so that the number of band pairs in the post-moult resilin equals the number of days the animal has lived since the moult.

It has been demonstrated that this band pattern is due to the light–temperature rhythm to which the animals are exposed when reared at normal day–night conditions[36,40]. Strongly fluorescent resilin is deposited during day conditions with high temperature (35°) and light, and less fluorescent resilin is produced under night conditions (25°) and darkness. Of the two factors involved temperature is by far the most important, light playing only a minor role. Strongly and homogeneously fluorescent resilin can be produced by rearing the animals in either darkness or light at constant high temperature.

Although the differences in fluorescence between the day- and night-layers can be seen clearly, the actual differences in intensity are quite small as demonstrated by microdensitometric scanning of negatives of photographs taken in the fluorescence microscope of sections of prealar arms[40]. This may explain why only small variations in the amounts of di- and trityrosine could be found when resilin was analysed from locusts grown under different external conditions[6].

A hypothesis has been suggested for the formation of the cross-links in resilin[6], based on the finding that di- and trityrosine can be formed artificially as cross-links in some proteins by oxidizing the tyrosine residues with hydrogen peroxide and a peroxidase. Free radicals are presumably formed as intermediates, and dityrosine is formed by the random pairing of such radicals. A peroxidase has been shown to be present in the resilin-producing epidermal cells but enzyme activity was also found to be present in the rest of the epidermis[41], and it is therefore not certain that the enzyme is involved in crosslinking resilin. A peroxidase has also been suggested to be involved in the formation of hard cuticle in insects, the evidence being that peroxidase-containing vesicles can be found in cuticle-forming cells, and these vesicles are apparently secreted into the cuticle[42].

The results so far obtained for resilin all indicate that in locusts resilin is formed by specialized epidermal cells which for about three weeks secrete a soluble protein which is immediately made insoluble by the formation of random cross-links between tyrosine residues in such a way that the final result is an insoluble isotropic material. The postulated soluble, non-crosslinked precursor of resilin has been called proresilin. It is not known whether this precursor is a single protein or a mixture of several, as it has not been possible to isolate the precursor, but determinations of the number and types of N-terminal amino acids present in resilin indicate that the protein chains have an average molecular weight of about 100 000 and that two different amino acids (serine and glutamic or aspartic acid) are present as N-terminal amino acids in approximately equal amounts[6]. This could indicate that two different proteins are involved in the formation of resilin, but better evidence is clearly needed to decide the question.

References p. 656

TABLE III

AMINO ACID COMPOSITION OF VARIOUS SAMPLES OF CUTICLE FROM MATURE FEMALE LOCUST (*S. gregaria*)

(Residues of amino acids per 100 residues)

	Hard cuticle							Soft cuticle		
	Ventral thorax	Lateral thorax	Abdominal terga	Abdominal sterna	Cornea of compound eye	Femur hind limb	Tibia hind limb	Abdominal intersegmental membrane	Membrane of coxal cavity	Neck membrane
Aspartic acid	3.4	3.9	4.7	4.9	3.4	3.5	3.6	8.4	9.5	7.7
Threonine	1.6	1.9	2.9	3.0	3.5	1.9	1.9	4.6	5.1	4.4
Serine	2.8	2.7	4.1	4.1	6.2	2.3	2.5	6.1	7.1	5.2
Glutamic acid	2.7	2.5	3.8	3.9	6.8	2.5	2.8	9.6	9.1	8.2
Proline	10.7	9.8	8.9	6.7	11.4	11.0	10.5	13.2	10.7	11.5
Glycine	11.9	10.6	12.4	8.1	7.7	10.5	10.9	18.0	20.5	19.8
Alanine	37.2	39.9	34.7	37.0	27.6	37.6	37.0	9.5	9.4	14.2
Valine	9.4	9.2	9.0	9.6	11.7	8.9	9.7	6.3	6.0	6.8
Cystine	—	—	—	—	—	—	—	—	—	—
Methionine	—	—	—	—	—	—	—	—	—	—
Isoleucine	4.0	4.0	3.5	3.9	3.2	4.1	4.1	4.5	4.5	4.9
Leucine	6.9	6.7	6.3	7.1	4.8	7.1	7.2	5.1	4.7	4.6
Tyrosine	3.1	2.2	2.4	3.3	5.5	3.5	2.8	2.2	1.7	1.7
Phenylalanine	0.9	0.8	0.9	1.1	1.4	0.9	0.8	3.0	3.3	2.8
Lysine	1.0	1.1	1.6	1.8	1.4	1.2	1.2	3.4	3.4	3.2
Histidine	2.1	2.1	2.0	2.4	2.3	2.3	2.5	1.8	1.4	1.5
Arginine	2.2	2.6	2.7	3.0	3.0	2.8	2.7	4.5	3.8	3.7
Sum of polar residues	18.9	19.0	24.2	26.4	32.1	20.0	20.0	40.6	41.1	35.6

The crosslinking of the soluble precursor(s) has been suggested[6] to take place just outside the cell-membrane with the enzyme either localized on the membrane or being secreted extracellularly so that the tyrosine residues in proresilin can become oxidized immediately after they have left the cells. The tyrosine radicals formed will then pair spontaneously and at random with each other giving rise to the three-dimensional chain-network present in fully formed resilin.

7. Resilin compared to other cuticular proteins in locusts

Apart from resilin, locusts contain two other main types of cuticle: the ordinary hard, sclerotized cuticle covering most of the body, and the soft flexible cuticle (arthrodial membrane) which interconnects the hard cuticle of the abdominal segments and of the limbs and which makes movements possible. The hard cuticle is nearly colourless except for a few instances where it is darkly coloured such as the mandibles, the tibial spines, and spots on the forewings. The soft cuticle is colourless.

All three types of cuticle are covered by a thin epicuticle containing protein and lipid but no chitin. The epicuticle is important in determining the permeability properties of the cuticle although it does not contribute significantly to the total mass of the cuticle[9]. The procuticle is located between the epicuticle and the epidermal cell layer and it constitutes the bulk of the cuticle. The procuticle can be divided into two layers, an outer exocuticle corresponding to the part of the procuticle deposited before moulting, and an endocuticle formed during the post-moult period. In all types of cuticle the procuticle consists of protein and chitin, and the differences between the types must be due partly to different proteins being involved and partly to the manner in which the proteins and the chitin are organized into molecular structures.

Since chitin is discussed in another chapter of this book I shall restrict myself to a comparison between the proteins in the different types of locust cuticle and between the ways in which they are being crosslinked.

The amino acid composition of various samples of hard and soft locust cuticle is given in Table III which can be compared with the composition of locust resilin in Table I. It will be seen that although there are some differences in composition between the various samples of hard cuticle these differences are minor compared to the difference between hard and soft cuticle. Alanine is the dominating amino acid in hard cuticle (28–40% of the total number of residues), whereas in soft cuticle and in resilin glycine is the most abundant amino acid (approximately 20 % and 40 %, respectively, of the total number of residues). It appears that the rule suggested for crickets[43], that hard cuticle contains more alanine than soft cuticle does, is also valid for locusts, although preliminary analyses of beetle cuticle show that this rule is not applicable to all species of insects.

The total amount of polar residues (aspartic acid, serine, threonine, glutamic acid, tyrosine, lysine, histidine and arginine) is significantly lower in hard cuticle than in either soft cuticle or resilin. The cornea from the compound eye has an intermediate composition with the content of alanine being high and polar residues being abundant. This cuticle is also less hard than cuticle from the legs or body wall although it is not flexible like the intersegmental membrane.

TABLE IV

AMINO ACID COMPOSITION OF FRACTIONS OBTAINED BY EXTRACTING HARD AND SOFT LOCUST CUTICLE WITH FORMAMIDE FOR 20 h

(Residues of amino acids per 100 residues)

	Femur cuticle		Intersegmental membrane	
	Residue	Extract	Residue	Extract
Aspartic acid	3.3	3.4	4.5	9.6
Threonine	1.7	1.6	3.7	4.5
Serine	2.4	1.8	6.3	5.6
Glutamic acid	2.2	2.7	6.1	10.7
Proline	11.0	11.4	11.9	11.9
Glycine	10.1	13.0	30.1	14.5
Alanine	37.2	35.9	7.4	12.7
Valine	9.5	9.3	8.5	6.5
½ Cystine	—	—	—	—
Methionine	—	—	—	—
Isoleucine	4.0	3.6	4.3	4.8
Leucine	7.4	7.7	6.3	4.7
Tyrosine	4.4	3.0	0.5	2.5
Phenylalanine	0.8	1.0	2.0	3.4
Lysine	0.9	1.2	2.9	2.8
Histidine	2.1	2.2	2.3	1.5
Arginine	2.8	2.3	3.5	4.2
Sum of polar residues	19.8	18.2	29.8	41.4

Table IV gives the amino acid composition of the extract and residue obtained by extracting hard and soft cuticle with formamide, which like urea solutions will dissolve hydrogen-bonded proteins. It can be seen that whereas the extract and residue from hard cuticle have nearly identical composition this is not the case for the soft cuticle, indicating that soft cuticle contains a mixture of proteins having different solubilities in form-amide. Disc electrophoresis of the extract from soft cuticle shows that several proteins are present here[44]. Whether more than one protein is also present in the residue and why the residue is insoluble in formamide is not known. It might be due to crosslinking or to bonds between proteins and chitin as suggested for larval cuticle by Hackmann[45].

Disc electrophoresis also shows that several proteins are present in the extract from hard femur cuticle[44]. Since extract and residue have identical composition the proteins present in the extract can be taken as representative of the proteins in the cuticle, indicating that something must have happened to the main part of the cuticular protein to make it insoluble. It is not only the exocuticle which is insoluble, as the amount extracted is much less than the total amount of either endo- or exocuticle present.

TABLE V

SOLUBILITY OF PROTEINS FROM LOCUST FEMUR CUTICLE IN FORMAMIDE

Samples were taken at various intervals after the final moult.

Interval since moulting	0 h	1 h	2 h	4 h	8 h	17 h	27 h	6 weeks
Protein extractable with formamide in percentage of total protein present	91.5	81.4	62.4	45.9	31.1	33.3	30.0	12.6

That an insolubilization of cuticular proteins takes place in locust can best be seen from the changes in extractability of the proteins taking place during the first hours after the final moult (Table V). There can be little doubt that some process occurs which makes the proteins less extractable, and at the same time the originally soft and pliable cuticle becomes hard and stiff.

This hardening and insolubilization of part of the cuticle after ecdysis is common to all insects, and in many insects it is accompanied by a darkening of the cuticle. It was suggested in 1940 by Pryor[46] that what happens is a tanning or crosslinking of the cuticular proteins caused by the reaction between free amino groups in the protein and some low molecular weight quinone formed by the oxidation of a diphenol. This hypothesis has never been seriously challenged, the main controversies being which diphenols are involved and how they are formed.

According to the best evidence now available N-acetyldopamine is the diphenol involved in the hardening of the cuticle in several insect species, among them the locust *Schistocerca gregaria*[47-49]. The main evidence for this is that injection of radioactive tyrosine, DOPA, dopamine, or N-acetyl-dopamine into insects at the right stage of development leads to the incorporation of the main part of the activity into the cuticle in an insoluble form. Enzymes which can transform tyrosine to DOPA, DOPA to dopamine, and dopamine to N-acetyldopamine are present, but the reaction sequence is only operative during periods of moulting. Outside these periods tyrosine is metabolized by another route involving degradation of the side-chain. From one insect (blowfly larvae) an enzyme has been isolated which will oxidize N-acetyldopamine to a quinone and which has no activity towards *ortho*-diphenols with acidic side-chains[50].

There is thus good support for the hypothesis that the exocuticle is being crosslinked by a process of bulk-tanning whereby a mass of proteins which have been secreted from the epidermal cells is rendered insoluble by the crosslinking action of a small reactive molecule, a quinone, the precursor of which (N-acetyldopamine) is being secreted from the cells into the protein. There is histochemical evidence[51] that a phenoloxidase is located in the epicuticle, so that the oxidation of the diphenol and therefore the tanning of the cuticle will take place from the outside inwards.

It has been suggested that the difference between dark and lightly coloured cuticle is due to differences in the amount of quinone being formed during the sclerotization[52]. When a large excess of quinone is present a dark cuticle should result, whereas a lightly coloured cuticle should be produced when the quinone is present in only slight excess.

Only little is known about the nature of the actual links formed between quinones and the proteins, as the isolation of any reaction products has not yet been reported. From model experiments[53] with quinones and soluble proteins and peptides it has been found that the most reactive groups in proteins are in order of reactivity: SH-groups, N-terminal amino groups, and ε-amino groups in lysine residues. Cuticular proteins contain little if any sulphur, so it is assumed that only amino groups are involved in the reaction. It has also been suggested[54] that the reactive species is not mono-meric quinone, but that the quinone may have to polymerize and condense in order to be an effective crosslinking agent. This may be one of the reasons why no cross-links have so far been isolated from sclerotized cuticle.

It is not known whether a similar crosslinking mechanism also operates in the endocuticle. The endocuticle is generally considered non-sclerotized, but the results reported here indicate that the proteins in the endocuticle are made insoluble. It would be interesting to know more about this process, because the endocuticle is deposited over a long period of time and therefore it does not seem plausible that it can be sclerotized by the sudden release of a small reactive molecule. Whether a slow continuous release of quinones can explain the findings remains to be investigated.

Although our knowledge of all types of cuticle is still insufficient to allow a detailed comparison, some of the main differences can be summarized. All three types of cuticle (hard, soft, and rubber-like) are characterized by a specific protein composition, and whereas only few proteins appear to be involved in the formation of rubber-like cuticle several are involved in the formation of both hard and soft cuticle. The protein chains in resilin are in random-coil configuration, and from analogy with hard and soft cuticles from other insects the proteins in locust hard and soft cuticle can be supposed to be in the β-configuration. Crosslinking in resilin takes place immediately after the secretion of the protein and goes on continuously during the formation of the ligaments. The crosslinking involves the oxidation of protein-bound tyrosine to free radicals which then pair at random to form a three-dimensional deformable network. Hard exocuticle is being crosslinked in bulk after the proteins have been deposited outside the cells. The crosslinking is being done by letting a low molecular weight quinone diffuse into the mass of protein where it reacts with available amino groups and probably also polymerizes. A large amount of the protein in endocuticle from hard cuticle and in soft cuticle is also insoluble and presumably crosslinked, but nothing is known about the processes which may be in-volved here.

ACKNOWLEDGEMENT

A grant from the Agricultural Research Council for the purchase of an automatic amino acid analyser is gratefully acknowledged.

References p. 656

REFERENCES

1   T. WEIS-FOGH, *J. Exptl. Biol.*, 37 (1960) 889.
2   K. BAILEY AND T. WEIS-FOGH, *Biochim. Biophys. Acta*, 48 (1961) 452.
3   S. O. ANDERSEN AND T. WEIS-FOGH, *Advan. Insect Physiol.*, 2 (1964) 1.
4   G. F. ELLIOTT, A. F. HUXLEY AND T. WEIS-FOGH, *J. Mol. Biol.*, 13 (1965) 791.
5   S. O. ANDERSEN, *Biochim. Biophys. Acta*, 69 (1963) 249.
6   S. O. ANDERSEN, *Acta Physiol. Scand.*, 66, Suppl. 263 (1966) 1.
7   S. O. ANDERSEN, *Biochim. Biophys. Acta*, 93 (1964) 213.
8   R. L. HILL, *Advan. Protein Chem.*, 20 (1965) 37.
9   V. B. WIGGLESWORTH, *The Principles of Insect Physiology*, 6th ed., Methuen, London, 1965, p. 30.
10 A. C. NEVILLE, *J. Insect Physiol.*, 9 (1963) 265.
11 M. JENSEN AND T. WEIS-FOGH, *Phil. Trans. Roy. Soc. (London)*, Ser. B, 245 (1962) 137.
12 T. WEIS-FOGH, personal communication.
13 Y. BOULIGAND, *Comp. Rend.*, 261 (1965) 3565.
14 A. C. NEVILLE, *Advan. Insect Physiol.*, 4 (1967) 213.
15 J. M. WITTEN, *J. Morphol.*, 127 (1969) 73.
16 H. C. BENNET-CLARK AND E. C. A. LUCEY, *J. Exptl. Biol.*, 47 (1967) 59.
17 C. NEVILLE AND M. ROTHSCHILD, *Proc. Roy. Entomol. Soc. (London)*, Ser. C, 32 (1967) 9.
18 A. SANNASI, *J. Georgia Entomol. Soc.*, 4 (1969) 31.
19 A. N. CLEMENTS AND S. A. POTTER, *J. Insect Physiol.*, 13 (1967) 1825.
20 B. L. GUPTA AND D. S. SMITH, *Tissue and Cell*, 1 (1969) 295.
21 P. L. MILLER, *J. Exptl. Biol.*, 37 (1960) 237.
22 A. SANNASI, *Experientia*, 26 (1970) 154.
23 F. LABELLA, F. KEELEY, S. VIVIAN AND D. THORNHILL, *Biochem. Biophys. Res. Commun.*, 26 (1967) 748.
24 F. W. KEELEY, F. LABELLA AND G. QUEEN, *Biochem. Biophys. Res. Commun.*, 34 (1969) 156.
25 S. KIMURA AND M. KUBOTA, *Bull. Japan. Soc. Sci. Fisheries*, 33 (1967) 430.
26 S. KIMURA AND M. KUBOTA, *J. Biochem.*, 65 (1969) 141.
27 M. IGNAC AND K. T. JOSEPH, *Leather Science*, 14 (1967) 213.
28 T. WEIS-FOGH, *J. Mol. Biol.*, 3 (1961) 648.
29 T. WEIS-FOGH, *J. Mol. Biol.*, 3 (1961) 520.
30 L. R. G. TRELOAR, *The Physics of Rubber Elasticity*, 2nd ed., Clarendon, Oxford, 1958.
31 A. J. GROSS AND I. W. SIZER, *J. Biol. Chem.*, 234 (1959) 1611.
32 S. O. ANDERSEN, unpublished observation.
33 S. S. LEHRER AND G. C. FASMAN, *Biochemistry*, 6 (1967) 757.
34 S. S. LEHRER, personal communication.
35 S. O. ANDERSEN AND B. KRISTENSEN, *Acta Physiol. Scand.*, 59, Suppl. 213 (1963) 15.
36 B. I. KRISTENSEN, *J. Insect Physiol.*, 12 (1966) 173.
37 B. I. KRISTENSEN, *J. Insect Physiol.*, 14 (1968) 1135.
38 B. GUPTA, personal communication.
39 A. C. NEVILLE, *J. Insect Physiol.*, 9 (1963) 177.
40 A. C. NEVILLE, *J. Cell Sci.*, 2 (1967) 273.
41 G. C. COLES, *J. Insect Physiol.*, 12 (1966) 679.
42 M. LOCKE, *Tissue and Cell*, 1 (1969) 555.

43   B. W. DE HASS, L. H. JOHNSON, J. H. PEPPER, E. HASTINGS AND G. L. BAKER, *Physiol. Zool.*, 30 (1957) 121.
44   J. WILLIS, personal communication.
45   R. H. HACKMAN, *Australian J. Biol. Sci.*, 13 (1960) 568.
46   M. G. M. PRYOR, *Proc. Roy. Soc. (London), Ser. B*, 128 (1940) 393.
47   P. KARLSON AND C. E. SEKERIS, *Nature*, 195 (1962) 183.
48   I. SCHLOSSBERGER-RAECKE AND P. KARLSON, *J. Insect Physiol.*, 10 (1964) 261.
49   P. KARLSON AND P. HERRLICH, *J. Insect Physiol.*, 11 (1965) 79.
50   P. KARLSON AND H. LIEBAU, *Z. Physiol. Chem.*, 326 (1961) 135.
51   R. DENNELL, *Proc. Roy. Soc. (London), Ser. B*, 134 (1947) 79.
52   H. S. MASON, *Nature*, 175 (1955) 771.
53   H. S. MASON, *Advan. Enzymol.*, 16 (1955) 105.
54   R. H. HACKMAN, in M. ROCKSTEIN (Ed.), *The Physiology of Insecta*, Vol. III, Academic Press, New York, 1964, p. 471.

Chapter XI

Elastin*

CARL FRANZBLAU**

Boston University School of Medicine, Boston, Mass. (U.S.A.)

1. Introduction

In the introductory paragraph to their treatise on elastic tissue in 1902, Richards and Gies[1] stated that, "Comprehension of function is dependent on knowledge of structure and composition. Elastic tissues have received little analytic attention. They have been overlooked by reason, apparently, of their seeming metabolic passivity and because they serve mainly mechanical functions." In 1970, the same introduction to a comprehensive review of elastin is still valid. It is true that to describe a structure–function relationship for a biological polymer such as elastin one must have a reasonable knowledge of its structure, *i.e.* primary, secondary and tertiary levels, if they exist, as well as a working knowledge of its function, which in the case of elastin is, as stated above, mechanical. Although it is also true that relatively few investigators have actually worked on the structure and function of elastin, much has been accomplished since the turn of the century.

It is important to mention at this point several pertinent reviews which have been published on the subject of elastin and it would be foolhardy for this author to include in his discussion many of the areas of elastin research already covered so well in previous articles. Those that must be mentioned include reviews by Partridge[2] in 1962, Ayer[3] in 1963, Seifter and Gallop[4] in 1966, Mandl[5] in 1961 and Piez[6] in 1968. What is intended in this

* Supported by grants from the National Institutes of Health (AM-07697) and the American Heart Association (68-777).
** Established Investigator of American Heart Association.

review is to focus primarily on the chemical crosslinks of elastin. In order to develop this more recent phase of elastin research, a reasonable but by no means exhaustive review of the elastin literature is necessary.

2. Isolation of elastin

Elastin is an insoluble protein found in most connective tissues in conjunction with other structural elements such as collagen and mucopolysaccharide. It is usually defined as that protein material which remains after all other connective tissue components have been removed. Essentially all methods of purification of elastin adhere to these two basic principles. The method by which non-elastin components are removed from a connective tissue has been really the only significant difference between the purification procedures used for elastin at the end of the 19th century and those that are used today.

According to Richards and Gies[1], Tilamus in 1840 is probably credited with describing the first purification of elastin from elastic tissues. The method employed cold-water extraction to remove traces of blood and inorganic matter, followed by dehydration with alcohol and ether. If, in addition, the residue was treated with boiling dilute acetic acid it was found to be free of sulfur. This material according to Tilamus was pure elastin. Müller[7] improved on the procedure of Tilamus by adding two additional steps which involved treatment with boiling dilute alkali (KOH) and treatment with cold mineral acid (dilute HCl).

Horbaczewski[8] in 1882 in his purification of elastin from cervical ligament introduced repeated extraction in boiling water. In this way he was able to convert collagen to gelatin and thereby remove it. Chittenden and Hart[9] pointed out the danger in the use of hot alkali and noted that a preparation of elastin differed in its sulfur content depending on whether or not it had been treated with hot alkali. Richards and Gies[1] probably were the first to note the relative ease with which elastin from *ligamentum nuchae* could be studied. These authors used cold-lime water instead of boiling KOH, followed by successive treatment of the residue with boiling water, boiling 10% acetic acid, and 5% HCl at room temperature. After removal of the HCl by repeated cold-water washes, the residue that remained was treated with alcohol and ether. Instead of the boiling water, the use of the autoclave was introduced in 1928 by Schneider and Hájek[10]. Purification of elastin from aortic tissue was accomplished with an 89% formic acid treatment. Hass[11] reported optimum condition for the formic acid purification procedure to be 45° for 72 h. For every 5 mg of dry tissue being purified, 1 ml of reagent was used. Under these conditions 90–95 % of the elastic tissue was recovered. Lowry et al.[12] employed alternatively autoclave procedures and hot 0.1 N NaOH treatment to determine the concentration of elastin and collagen in several tissues. Lansing et al.[13] have described a purification of elastin from aortic tissue which also includes 0.1 N NaOH. After washing with 0.9% NaCl the authors heated the aortic material in 0.1 N NaOH for 45 min at 95°; the alkali was removed by washing with water and the residue finally treated with alcohol and ether to effect the removal of lipid and water from the purified elastin. Partridge et al.[14] utilized an exhaustive autoclave procedure to essentially obtain pure elastin from *ligamentum nuchae* of the ox. The latter is the most commonly used source of tissue for the preparation and purification of elastin. One is usually required to autoclave the previously milled and saline-extracted material for successive periods of 1 h until no further protein appears in the supernatant. This treatment, although harsh, yields a preparation of elastin with an amino acid composition which remains constant upon further treatment in the autoclave. Elastin purification from aortic tissue cannot be accomplished by autoclaving alone. One is still required to treat the washed aortic material with cold alkali. As pointed out so frequently, exposure to hot alkali and/or exhaustive autoclaving may lead to extensive peptide-bond cleavage which may not be readily detected in insoluble elastin preparations.

Attempts have been made to avoid the use of such drastic treatments. Hospelhorn and Fitzpatrick[15] first proposed the use of enzymes to purify elastin. According to their procedure, elastic tissue is extracted with 1 N NaCl, dried, defatted and then digested alternately with trypsin and collagenase. The product that results has an amino acid composition different from alkali-extracted elastin. More recently, Miller and Fullmer[16] described a purification procedure for elastin from chick aorta. The tissue is extracted at neutral pH and 5° for successive 48-h periods with 3% NaH_2PO_4, 25% KCl and 5 M guanidine–HCl. These treatments were followed by incubation of the residue with a commercially available purified collagenase, 1 mg of enzyme and 10 mg of insoluble substrate. Water washing followed by a repetition of the guanidine extraction and collagenase treatment yielded a purified elastin residue which was dried by lyophilization. Comparison of the elastin prepared in this manner with that prepared by the Lansing et al.[13] procedure revealed some slight differences in amino acid content.

References p. 709

Steven and Jackson[17] treated homogenized tissue with crude bacterial α-amylase to remove most of the collagen as a dispersion in acetic acid. The residual collagen is then removed by digestion with collagenase for 4 h at pH 8.0 in 0.1 M $CaCl_2$ using an enzyme–substrate ratio of approximately 1/50 (w/w). However, elastins prepared in this way showed a slightly higher hydrophilic amino acid residue content. When fetal bovine aorta and *ligamentum nuchae* were first treated with α-amylase before hot alkali and collagenase treatment, the combined enzyme and alkali treatments yielded insoluble fetal elastin which was purer than that produced by hot alkali alone.

Ross and Bornstein[18] recently reported the isolation of fetal bovine *ligamentum nuchae* elastin by extraction with 5 M guanidine followed by digestion with collagenase. The preparation of elastic fibers obtained by this method showed two morphologically different constituents under the electron microscope, a central amorphous region and the surrounding myofibrillar component. The latter was removed by proteolytic enzymes or by reduction of disulfide bonds with dithioerythritol in 5 M guanidine. The amino acid composition of the amorphous component was identical with that previously described for elastin.

3. Mechanical properties of elastin

Studies on the elastic properties of elastin have been hampered by the problems inherent in the purification procedures mentioned above. Burton[19] in his review on the properties of blood vessels explains in detail and with great clarity the several parameters which may be measured.

Elasticity is defined as the property of materials which enables them to resist deformation by the development of a resisting force or "tension". All coefficients of elasticity are defined as the ratio of this resisting force to the measure of the deformation produced. Accordingly, Burton points out that a material which possesses "high elasticity" is one which resists deformation, while a material of "low elasticity" does not resist deformation and a small force to the latter will produce large deformation. Thus, glass or steel has a much higher elasticity than does rubber.

Another parameter pointed out by Burton is related to the degree of deformation or stretch that is possible before there is an irreversible change. This characteristic encompasses "tensile strength", "breaking strength" and "yield point". "Tensile strength" refers to a load which produces an irreversible change, while "breaking strength" is the load required to produce actual rupture. The "maximum extension", based on percentage of the original length before the yield point is reached, is also of interest when describing elasticity.

When discussing biological fibers and their elastic properties, one almost certainly encounters Young's modulus, which is defined as

$$F = \frac{Y \, \Delta l}{l_0} \cdot A$$

where F is the force in dynes exerted on the material, Δl is the extension produced beyond the nonstretched length l_0, and A is the cross-sectional area in square centimeters. Y is given in dynes per square centimeter for 100% elongation. These various parameters are given in Table I taken from the review of Burton[19].

TABLE I

ELASTIC PROPERTIES OF ELASTIN, COLLAGEN, RUBBER AND STEEL[a]

Substance	Young's modulus $(dynes/cm^2/100\% \, el.)$	Tensile strength $(dynes/cm^2)$	Maximum extension (%)
Elastic fibers	$3 \cdot 10^6$	$1 \cdot 10^7$	100
Collagenous fibers	$1 \cdot 10^9$	very great	50
Rubber	$4 \cdot 10^7$	$2 \cdot 10^8$	600
Steel	$2 \cdot 10^{12}$	$2 \cdot 10^{12}$	0.1

[a] From Burton[19].

Burton pointed out that the elastic constants of elastin (elastic fibers) are quite extensible compared to most substances, actually more extensible than rubber. He compared the extensibility of a rubber band having a cross-sectional area of 0.04 cm^2 and a 50-g load attached to it with an elastic fiber from aortic tissues of the same cross-sectional area and a 50-g load attached to it. The rubber band stretched approximately 7% while the elastic fiber stretched 40%. In summary we can say that elastin possesses a high extensibility combined with a low modulus which is very similar to the elastic properties of rubber.

It had been shown by Wohlisch et al.[20] and Meyer and Ferri[21] that the stress–strain curve for elastin, like that of rubber, swings upward at high extensions. However, the rise is more abrupt and it occurs at somewhat lower elongations than that of rubber. To explain this Wohlisch suggested that in the case of elastin, crystallization occurs on stretching. However Astbury[22], examining the X-ray diffraction patterns of stretched elastin, found no evidence for such crystallization.

Hoeve and Flory[23] reexamined the elastic properties of unpurified ox *ligamentum nuchae* and concluded that no crystallization occurs when elastin is stretched. Furthermore, they concluded that the stress–strain curve is explained by the morphology of native elastin; the abrupt rise in stress at high elongations is attributed to straightening out of the initially curled fibers of collagen which are associated with native elastin in the unpurified ligament. The erroneous results of earlier workers were attributed to the fact that they did not take into account the de-swelling of elastin in water with elevated temperatures. The data of Hoeve and Flory were obtained in 30% glycerol where the degree of elastin swelling in this solvent is independent of temperature. However, Ayer[3] in his extensive review suggested that the non-Gaussian increase of stress at high extensions is not caused by fully extended collagen fibers as suggested by Hoeve and Flory, since the same effect is seen in collagen-free elastin.

Mukherjee[24] reexamined the stress–strain properties of purified *ligamentum nuchae* elastin. The specimens studied contained two kinds of void spaces. "Macrovoids" were due to the removal of collagen and mucopolysaccharides and "microvoids" was the term assigned to those spaces between the peptide chains within the bulk elastin molecular regions.

The stress–strain properties of the purified elastin immersed in deionized water exhibit three distinct regions. At low load regions (0–200 g/cm^2) there is an initial slack in the non-woven fabric-like structure which results in an initial low modulus, linear stress–strain curve. At intermediate loads (200–1000 g/cm^2) the stress–strain curve is also linear but with a higher modulus, and it was suggested that in this region the stress is now carried by the peptide chains on a molecular level. At both low and intermediate loads, the stress–strain behaviour is reversible. At higher loads (1000 g/cm^2) the fibers begin to slip irreversibly and to rupture.

By application of the kinetic theory of rubber elasticity[25] the slopes of the stress–strain curves are related to the molecular weight (Mc) between cross-links. Data were obtained for elastin preparations immersed in both aqueous and non-aqueous solutions. The Mc value of 3400 obtained in deionized water increases with the increasing concentrations of formamide, reaching a maximum of 7900 at a formamide: H$_2$O concentration of 44% (v/v). The increase in Mc was shown to be irreversible. This was explained by the fact that the peptide chains in elastin assume a new equilibrium conformation. Similar findings were obtained in ethanol and butanol solutions. These results are compatible with those of Robert and Poullain[26] who examined the effect of alcohols on the rate of hydrolysis of elastin in alkali.

4. Morphology and distribution

Early work on elastic tissue was based primarily on the unique staining properties of elastic fibers. These histological properties, together with an ability to stretch when wet, rendered elastic tissue relatively easy to identify. Orcein and Weigert's resorcin–fuchsin, the most widely used dyes, have been shown to stain elastin selectively. Other histological stains which have been used include Verhoeff's hematoxylin, Nile Blue Sulfate and Sudan Black[27]. For a thorough review on the histochemical determination of elastin the reader is referred to the review of Dempsey and Lansing[28].

Available evidence indicates that there is a sparsity of elastin in skin, tendon and loose connective tissue. It has been reported that elastin comprises only 2–5 % of the dry weight of skin[29]. On the other hand, relatively large amounts of elastin are found in ligaments and large blood-vessel walls. The content of elastin in aorta[30,31] is 30–57 % and in *ligamentum nuchae*[4] 78–80 %.

Morphologically, the fibers of elastin are not of one single type. Briefly, one can say that in aorta the *tunica media* accounts for the majority of the elastic tissue of the vessel. The fiber takes the form of 50–60 concentric elastic membranes about 2.5 μ thick which are separated by layers of tissue 6–18 μ thick, consisting of collagenous and elastic fibers, lipid material, smooth muscle cells, and fibroblasts embedded in ground substance. In *ligamentum nuchae* of the ox the elastic fibers are thick and under the light microscope preparations of this elastin appear as short, smooth rodlike fibers of almost circular cross-section. The thickness of such fibers appears to be uniform and gives a measure of 6.7 μ for the mean value of the diameter.

Electron micrographs of elastic fibers from a variety of sources reveal no ordered structure[32]. Because of the insoluble nature of elastin, no information concerning the size or shape of its molecule is available. Wide-angle X-ray diffraction data reveal no axial periodicity and, at present, such information suggests that elastin is a three-dimensional network of randomly coiled polypeptide chains joined by covalent crosslinks[33].

Of the more pronounced characteristics of elastin, its yellow color and strong blue-white fluorescence have been known for many years. Although the compound or compounds responsible for these properties have still not been completely characterized, there is general agreement that the fluorescent chromophore is firmly bound to the polypeptide backbone. This has been substantiated by studies on soluble fragments of elastin by LaBella[34], Kärkelä and Kulonen[35], Loomeijer[36], and Sinex[37], and their collaborators.

TABLE II

AMINO ACID COMPOSITIONS OF VARIOUS ELASTINS AND COLLAGEN[a]

Amino acid	Elastin, monkey aorta[b]	Elastin, human skin[c]	Elastin, bovine Lig. Nuchae[d]	Collagen, rat skin[e]
Hydroxyproline	—	6.4	7.1	92.0
Aspartic acid	3.2	2.8	7.3	46.0
Threonine	4.3	4.8	10.1	19.6
Serine	11.2	5.6	9.9	43.0
Glutamic acid	15.6	15.0	17.4	71.0
Proline	98.0	149.4	125.4	121.0
Glycine	354.6	330.9	316.2	331.0
Alanine	248.7	249.4	213.3	106.0
Valine	136.1	105.7	134.0	24.0
Methionine	—	—	—	7.8
Isoleucine	21.4	19.9	26.6	10.8
Leucine	58.6	54.1	64.7	23.8
Tyrosine	9.7	16.8	6.1	2.4
Phenylalanine	20.1	20.3	33.6	11.3
Isodesmosine	1.3	1.2	1.1	0
Desmosine	1.7	1.6	1.7	0
Hydroxylysine	0	0	0	92.0
Lysine	3.5	3.8	3.6	28.1
Histidine	—	—	0.5	4.9
Arginine	5.5	3.6	6.6	51.0
Lysinonorleucine	present	present	0.9	—

[a] Residues per 1000 total amino acid residues.
[b] Cannon (unpublished results).
[c] Varadi and Hall[29].
[d] Franzblau et al.[153].
[e] Piez et al.[154].

5. Amino acid composition

Regardless of the method of preparation, the amino acid composition of purified elastin is unique. As in collagen, one-third of the amino acid residues in elastin are glycine and one-ninth of the residues are proline. In contrast to collagen, elastin contains very little hydroxyproline, no hydroxylysine and a preponderance of the non-polar amino acids, alanine, valine, leucine and isoleucine. There are very few residues of aspartic acid, glutamic acid, lysine or arginine. As shown in Table II, the amino acid composition of elastin from aorta, *ligamentum nuchae*, and skin are similar to one another but considerably different from rat-skin collagen.

The amino acid which appears to have the greatest variability when examining preparations of elastin from several species or from different organs in the same species is tyrosine. The latter varies from 10 residues per 1000 residues in bovine *ligamentum nuchae*, to approximately 40 residues per 1000 residues in elastin obtained from the swim bladder of the carp[38]. Human aortic elastin, for example, contains 25 residues of tyrosine per 1000 residues. Of special interest is the unusually high content of valine residues in elastin. Practically nothing is known about the sequential arrangement of the amino acids in elastin. From an examination of the amino acid content, it is tempting to suggest that, as in collagen, elastin contains a repeating sequence. Because bacterial collagenase has little or no effect on elastin, such a repeating sequence, if present, cannot be similar to that found in collagen (*e.g.* –Gly–Pro–X–). It is of interest to note that alkaline hydrolysates of elastin contain a considerable amount of the dipeptide valylproline[39]. Most recently, Keller and Mandl[40] have isolated two other dipeptides in relatively high concentration, valyl valine and leucyl valine. Such valyl-containing peptides are known to be relatively resistant to acid and alkaline hydrolysis.

6. Quantitative estimation

Lowry *et al.*[12] determined elastin gravimetrically by weighing the dried residue from dilute alkali extraction at room temperature. Collagen was converted to gelatin by autoclaving with water after alkali treatment. This procedure was modified by Neuman and Logan[30] to determine hydroxyproline in the autoclaved extract for collagen and hydroxyproline in the residue for elastin.

Lansing *et al.*[13] showed that elastin could be prepared by heating whole defatted tissue with 0.1 N NaOH at 98° for at least 45 min, after which the weight of the residue was constant up to 60 min. This method is fast and convenient for routine examination of elastin in tissues with a high content of elastic fibers but is less applicable for tissues with higher muscle or other cell content. The weighed residue of elastin has to be corrected for ash since calcium is present in notable amounts in the residue after alkali treatment. Correction for moisture also has to be made. As an alternate method, the total nitrogen of the residue after acid hydrolysis can be estimated.

Scarselli[41] has suggested a specific method for the estimation of elastin based on the affinity of elastin for the dye orcein. Purified elastin is treated with orcein in acidic ethanol for 24 h. After the excess orcein is removed, the tissue with bound orcein is digested with elastase to dissolve the orcein–elastin complex. The latter is measured by colorimetry at 590 mμ. The validity of this method is still in question because of the findings of Engle and Dempsey[42] and Brolin and Hassler[43] that some component other than elastin may be stained and that orcein itself is a mixture of dyes only two of which are specific for elastin.

Jackson and Cleary[44] believe that with the data available at present, it is safe to conclude that elastins from different tissues and different mammalian species represent a single protein or group of proteins with a characteristic amino acid composition. This amino acid analysis serves as the standard for purity of the elastin preparation. Reservations must be held though with the tissue under study, the age of the animal from which it was obtained and the method of isolation.

Since it was found that the amino acids desmosine and isodesmosine are present only in elastin (to be discussed), their determination in tissues containing elastin can assume the same role as the determination of collagen on the basis of its hydroxyproline content. The determination of elastin on the basis of desmosines may even be more specific since relatively high amounts of hydroxyproline have been found in elastin[45–48] making the calculation of collagen content from hydroxyproline in tissues rich in elastin erroneous.

Anwar[49] reported that for quantitation experiments, the use of 0.38 N citrate buffer, pH 4.26 at 50°, gave well-separated peaks of desmosine and isodesmosine on the medium-length column (50 cm) of the Spinco amino acid analyzer from an acid hydrolysate of elastin. For isolation and characterization of these amino acids, separation on a 50-cm column with a 0.2 N citrate buffer, pH 4.45, was found to be more suitable. The desmosine and isodesmosine fractions were collected and desalted on a 15-cm column of Dowex 50X8 resin.

Ledvina and Bartoš[50] proposed a method whereby the elastin hydrolysate is applied to a Sephadex G-15 column, the fastest fraction then separated and passed through a Dowex 50WX4 column. The desmosines region is then collected and analyzed by UV absorption at 274 mμ.

7. Solubilization of elastin

(a) Enzymatic

The digestion of elastin by proteolytic enzymes has been examined in many laboratories. In 1949 Baló and Banga[51] isolated from pancreatic tissue an enzyme capable of solubilizing elastin. This enzyme, which was subsequently crystallized, was thought to be specific for elastin and was therefore designated an elastase (EC 3.4.4.7). Although this elastase does indeed dissolve elastin, it has been shown to attack other proteins as well. Naughton and Sanger[52] found that a highly purified preparation of elastase was capable of hydrolyzing the A and B chains of insulin. Analyses of the peptides obtained indicated that elastase catalyzed the hydrolysis of peptide bonds involving the carboxyl groups of neutral amino acids such as leucine and valine. Most recently, Narayanan and Anwar[53] reexamined the specificity of pancreatic elastase and found that the preparation of Naughton and Sanger[52] was not homogeneous. Further purification of elastase led to the conclusion that the specificity was much narrower than that which had been previously reported. Examination of synthetic peptides with a highly purified preparation of elastase[54] indicated a strong specificity for alanyl peptide bonds. It is also of interest that elastase has certain esterolytic activity and very simple procedures can be employed to detect elastase activity in various tissue preparations[55]. Grant and Robbins[56,57] showed that elastase is secreted in the pancreatic juice as an inactive zymogen, "proelastase", which can be activated by trypsin (EC 3.4.4.4) or enterokinase (EC 3.4.4.8). Lamy et al.[58] have presented evidence of a purified fraction of pancreatic elastase which is capable of hydrolyzing only elastin. If the latter results are substantiated, the availability of a specific elastase would be an invaluable probe in investigating the structure of elastin. For a more complete discussion of earlier work on elastase the reader is referred to the review of Mandl[5].

Among other mammalian enzymes examined for elastolytic activity, trypsin and chymotrypsin do not attack elastin. Studies with pepsin have led to conflicting results. Some laboratories reported elastolysis although others

were unable to demonstrate activity. Partridge[2] pointed out that these results may be attributed to differences in the preparations of elastin used in
the latter studies. Of the nonmammalian enzymes possessing elastolytic activity, pronase, papain (EC 3.4.4.10), ficin (EC 3.4.4.12), bromelain (EC
3.4.4.24) and Nagarse are included[2,59]. These enzymes are generally considered non-specific, being capable of hydrolyzing peptide bonds between
many different amino acid residues.

(b) Chemical

Since elastin is an insoluble polymer, its solubilization has been achieved
only by the disruption of peptide bonds, either by the action of the proteolytic
enzymes discussed above or by the use of weak acids such as oxalic acid.
By successive 1-h treatments of elastin from bovine *ligamentum nuchae* with
0.25 N oxalic acid at 100°, Partridge *et al.*[14] were able to solubilize bovine
elastin. This treatment resulted in the formation of two soluble components
designated α-elastin and β-elastin. α-Elastin was found to be polydisperse
with an average molecular weight between 60000 and 84000 while the β-
elastin was monodisperse and had an average molecular weight of approximately 5500. Examination of the amino-terminal residues of these two fractions by the use of fluorodinitrobenzene (FDNB) suggested that the α-protein
contained approximately 17 chains with an average of 35 amino acid residues per chain, while the β-protein was comprised of two chains with an
average of 27 amino acids per chain. It was postulated from these data that
a significant number of crosslinks must be present in elastin. This in itself is
not surprising since it had been pointed out earlier that a rubber-like elastomer
must contain a reasonable number of crosslinks in its structure[60]. These
crosslinks are necessary to impart a certain amount of restriction to the
elastomer so that while stretching under stress, one chain is not able to slip
completely past another. This preserves the ability of the elastomer to return
to its orginal unstretched shape when the stress is released.

Ioffe and Sorokin[61] used copper sulfate and 0.4 N barium hydroxide at
37° for 60 h to solubilize elastin. Hall and Czerkawski[62] have reported
the preparation of similar α-elastins by digesting elastin with purified pancreatic elastase in the presence of sodium dodecyl sulfate or by treatment
with hot ethanolic–HCl or 40% urea. These latter α-elastin preparations as
well as that of Partridge *et al.*[14] are all capable of forming a coacervate at
elevated temperatures. The phenomenon is pH-dependent, the optimum

however varies with the solubilization procedure. For example, the oxalic acid-prepared α-elastin yields a coacervate optimally at pH 5.5, while an α-elastin prepared[14] with ethanolic–HCl forms a coacervate optimally at pH 9.0. The latter when treated with mild alkali at room temperature resulted in a shift of the maximum[14] to pH 4.0. It will be seen later that this ability to form a coacervate from soluble elastin preparation was an important tool in the isolation of a soluble precursor elastin.

Robert and Poullain[63] have found that elastin can be solubilized in 0.1 N KOH in 80% ethanol at room temperature. This solubilized preparation has been designated k-elastin.

The soluble protein from a hydrolysis of bovine *ligamentum nuchae* elastin by pancreatic elastase was fractionated by LaBella[34] by the addition of trichloroacetic acid. The insoluble fraction, which resembled α-elastin, was produced in a small amount early in the incubation period and more rapidly towards the end. This material, which had precipitated to the extent of 80% when solubilization was complete, was yellow, Schiff-positive, nondialyzable and could be degraded further by elastase. This rate of production was the exact opposite of the trichloroacetic acid-soluble fraction which resembled β-elastin. In addition, the latter was colorless, Schiff-negative, had twice the fluorescence of the insoluble fraction, and could not be degraded further by the enzyme. The course of hydrolysis was thus similar to that with organic acids.

8. Nature of the crosslinkages

(a) Desmosine and isodesmosine

The amino acid analyses in Table II clearly indicate that crosslinking of elastin cannot be explained entirely by the presence of disulfide bridges (since there are very few) and suggest that some other type of crosslink must be present. The ultraviolet absorption of α-elastin provided the first clue to the nature of these linkages[14]. It was found that the total absorption of α-elastin at 275 mμ could not be accounted for by the known contents of tyrosine and phenylalanine in elastin, and the protein contains few or no tryptophan residues.

An explanation for this observation was provided by Partridge *et al.*[64]. Digestion of elastin with elastase followed by treatment with leucine amino-peptidase (EC 3.4.1.1) and carboxypeptidase (EC 3.4.2.1, 2) yielded, among other products, a peptide which contained an unidentified ultraviolet-absorbing substance. This peptide was shown to contain two carboxyl terminal residues and two α-amino groups as determined by titration and end-group analysis. Dinitrophenylation yielded DNP–alanine and a DNP derivative of the ultraviolet-absorbing compound; the latter was not released by acid hydrolysis. However, hydrolysis of the nondinitrophenylated peptide by

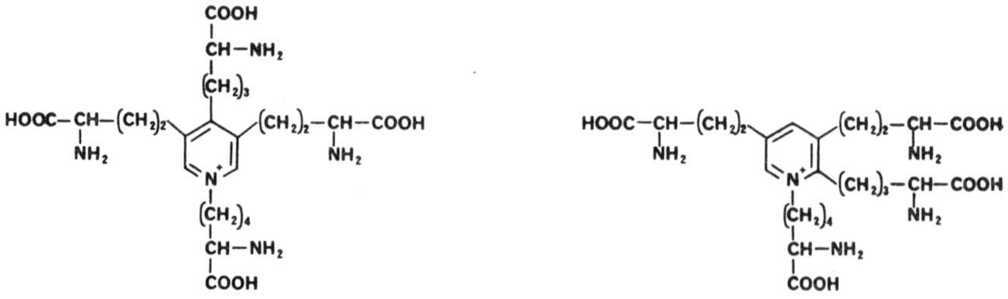

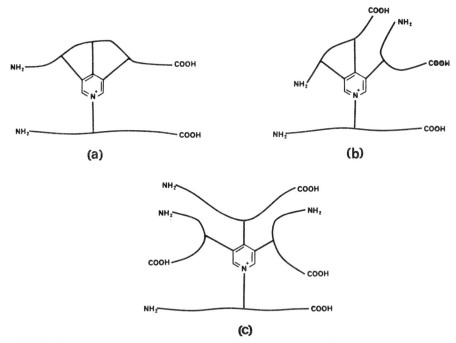

Fig. 1. Structure of desmosine and isodesmosine

(a)   **(b)**

(c)

Fig. 2. Schematic illustration of possible elastin crosslinks. One desmosine moiety may crosslink (a) two peptide chains, (b) three peptide chains, or (c) four peptide chains.

strong acid yielded the compound in the free form. Subsequent chemical analyses of hydrolysates of elastin revealed the presence of two polyfunctional amino acids containing a pyridine nucleus alkylated in four positions, including the ring nitrogen. These two compounds were found to be isomers differing only in position of ring substitution[65]. The postulated structures of these newly described amino acids, designated by Thomas *et al.*[65] as "desmosine" and "isodesmosine" respectively are shown in Fig. 1.

On the basis of their structure and the nature of the peptides in which they were contained, Partridge proposed that desmosine and isodesmosine are involved in the crosslinking of elastin. The number of peptide chains crosslinked by these unusual amino acids is still unknown. As illustrated in Fig. 2, two, three, or four chains could be crosslinked by one desmosine or isodesmosine molecule.

The ultraviolet absorption of these two pyridinium compounds compared well with the ultraviolet absorption of 1,2,3,5-tetramethylpyridine and 1,3,4,5-tetramethylpyridine. Ferricyanide oxidation led to the production of significant amounts of lysine and NMR data confirmed the correct positions of ring substitution. The empirical formula for both desmosine and isodesmosine was found to be $C_{24}H_{39}N_5O_8$. That the desmosines were present in several different elastin preparations was shown by Anwar[49] among others.

(b) Lysinonorleucine

To gain more insight into the nature of the crosslinking in elastin, Franzblau *et al.*[66] attempted to isolate, from various proteolytic digests of elastin, peptides enriched in desmosine and isodesmosine. In the course of this study, it was observed that such peptides could be obtained but that they contained yet another undescribed amino acid, which was initially designated as X_4.

This amino acid was isolated from a total acid hydrolysate of elastin by usual chromatographic techniques. In contrast to the desmosines, the purified material did not absorb ultraviolet light over the range 220–340 mμ. The infrared spectrum was typical of an amino acid. Dinitrophenylation studies, potentiometric titrations in the presence of formaldehyde, NMR spectra, and electrophoretic mobility of X_4 suggested the presence of two primary and one secondary amino groups. These data, together with determination of α-amino groups by a manometric ninhydrin method, total amino groups by a nitrous acid method, total nitrogen and total carbon led to formulation of the structure shown in Fig. 3.

Proof of structure of this compound was confirmed by synthesis. Oxidation of α-acetyllysine with *N*-bromosuccinimide yielded the *N*-acetyl derivative of α-aminoadipic-δ-semialdehyde. Treatment of this compound with an equivalent of unreacted α-acetyllysine followed by reduction either with hydrogen over palladium or with sodium borohydride gave a product which, when

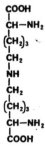

Fig. 3. Structure of lysinonorleucine.

hydrolyzed in 6 *N* HCl, was identical with the naturally occurring amino acid as studied by electrophoresis and by chromatography on ion-exchange resins. This compound, isolated from elastin, was treated with snake venom L-amino acid oxidase (EC 1.4.3.2), and was found to yield two moles of ammonia for each mole of amino acid. Thus, both α-carbons appear to be in the L-configuration. It was concluded that the newly isolated amino acid is N^{ε}-(5-amino-5-carboxypentyl)lysine, and the authors suggested the trivial name, lysinonorleucine.

More recently a much more efficient method for the synthesis of lysinonorleucine has been employed[67]. The procedure is a modification of that used by Speck *et al.*[68] for the synthesis of N^{ε}-glyceryllysine. 5-δ-Bromobutylhydantoin is refluxed with N^{α}-benzoyllysine ethyl ester in tetrahydrofuran; the reaction mixture is concentrated and then treated with alkali to hydrolyze the product. Lysinonorleucine is obtained in 55% yield. The ethyl ester was prepared, and its mass spectrum confirmed the structure given above for lysinonorleucine. The polyfunctional structure of lysinonorleucine suggests that it, like the desmosines, also serves a crosslinking function in elastin (see Fig. 4).

The discovery of these newly described amino acids provides true chemical markers for elastin in the sense that hydroxylysine and hydroxyproline are "labels" for collagen. On the basis of accumulated data, elastin of bovine *ligamentum nuchae* contains, per 1000 total amino acid residues, an average of 1.4, 1.0 and 0.9 residues of desmosine, isodesmosine, and lysinonorleucine, respectively. The values for desmosine and isodesmosine are surprisingly constant among preparations of mature elastin obtained from several sources. The lysinonorleucine content, however, appears to exhibit some variation.

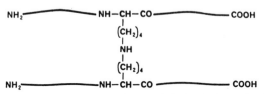

Fig. 4. Schematic illustration of possible elastin crosslink. One lysinonorleucine moiety may crosslink two peptide chains.

(c) *Biosynthesis of the crosslinkages*

As mentioned above, treatment of desmosine or isodesmosine with a mild oxidizing agent such as alkaline ferricyanide results in formation of a significant amount of lysine, which suggested to Partridge *et al.*[69] that the biosynthesis of the desmosines could be from lysine. Independent studies by Miller *et al.*[70] on chick embryo aorta and by Partridge *et al.*[71] on rat aorta indeed have confirmed that lysine is the precursor of both desmosine and isodesmosine. Aortic elastin preparations obtained from chick embryos at various times during development were examined for contents of desmosine, isodesmosine and lysine. The lysine content was relatively large in young embryos and decreased in the older embryos. The decrease in lysine was accompanied by a measurable increase in both desmosine and isodesmosine. The elastin in these experiments[69-71] was prepared by treating the dried, fat-free samples of ascending aorta with 0.1 N NaOH at 98° for 1 h according to the method of Lansing *et al.*[13]. Miller *et al.*[70] found that complete release of desmosine, isodesmosine and valine was not achieved unless 6 N HCl hydrolysis was carried out for 72 h. They also indicated that the ninhydrin reaction used in their amino acid analyses yielded a color equivalent for desmosine and isodesmosine equal to 3.68 times that of leucine under the same conditions. The data of Miller *et al.* are shown in Table III. Rather than expressing the data as residues of desmosine per 1000 residues of amino acid, several investigators have preferred to use a "quarter-desmosine" value. This

TABLE III

LYSINE AND QUARTER-DESMOSINE CONTENT OF ELASTIN FROM AORTAS OF CHICKENS OF DIFFERENT AGES[a]

	12-day embryo	16-day embryo	20-day embryo	3-week chick	4-week chick	1-year chicken
Lysine	5.7	4.6	3.9	3.6	3.5	1.6
Desmosine/4	4.3	5.7	6.7	6.8	7.1	10.9

[a] From Miller et al.[70]. Data expressed as residues per 1000 residues.

is based on the fact that there are four α-amino, α-carboxyl groups in each molecule of desmosine.

Other studies on the biosynthesis of the desmosine and isodesmosine involved the incorporation of [^{14}C]lysine either in the intact animal or in tissue culture systems. In the latter, a 24-h pulse of uniformly labeled [^{14}C]-lysine was usually supplied to cultures of chick embryo aortas and the specific activity of the desmosines was determined with time after pulse. Miller et al.[70] removed approximately 3 mm of the ascending aorta from 16-day-old chick embryos. These were attached to a Millipore filter by means of a fibrin clot and placed individually in test tubes. For each time period 25–50 aortas were used. The culture medium employed was Medium 199 (with Hank's salts) containing in addition to penicillin and streptomycin, approximately 0.1–0.5 μC/ml of uniformly labeled [^{14}C]lysine. Those aortas which had been treated with radioactive lysine for one day and then grown in medium containing no lysine for an additional 11 days (the medium was changed every other day throughout the chase period) gave the following results after isolation and hydrolysis of the elastin. Lysine had a specific activity of 73 700 counts/μmole and the quarter-desmosines (desmosine and isodesmosine) had a specific activity of 78 100 counts/μmole. This suggested that four lysine residues are incorporated into one desmosine or isodesmosine moiety, since there are 24 carbon atoms in a molecule of desmosine and 6 carbon atoms in a molecule of lysine. Similar data were obtained by injecting [^{14}C]lysine into the intact animal. Partridge et al.[71] studied the incorporation of lysine into desmosine employing Sprague-Dawley rats. 0.1 mC of radioactive lysine was injected into a rat in 30 equal doses at approximately 8-h intervals for 10 days. The animals were sacrificed about 12 h after the last injection and the elastin from aorta was prepared according to the procedure of Lansing et al.[13]. The resulting protein was hydrolyzed in HCl and the ratio of specific activity in the desmosines (not quarter-desmosines) to the specific activity in lysine was found to be greater than 2.0. The possibility that four lysines were incorporated into the desmosines was not excluded by the authors since in the maturing elastic fiber there may be a considerable lag in time before completion of the full complement of desmosine crosslinks.

Additional evidence that four lysines are incorporated into one desmosine or isodesmosine moiety was obtained by Anwar and Oda[72]. These authors had previously shown[73] that treatment of isodesmosine with 10 N NaOH at 110° for 16 h resulted in the formation of lysine in reasonable yield. Isodesmosine gave 73% of the theoretical lysine value. On the other hand, desmosine yielded 41% of its theoretical value of lysine by treatment with 6 N NaOH at 110° for 70 h. Radioactive desmosine was obtained from the ascending aortas of 13–14-day-old chick embryos which had been grown in tissue culture in the presence of uniformly labeled [^{14}C]lysine. Both desmosine and isodesmosine were isolated by column chromatography and then subjected to alkaline degradation. The ratio of the specific activity of the entire desmosine or isodesmosine molecule to the specific activity of the lysine obtained after degradation was determined. In both cases the ratio was found to be approximately 4.0. In addition, the isolated radioactive desmosines were treated with ninhydrin and the ratio of the total activity of the molecule to the total activity of the CO_2 recovered after such a procedure was found to be approximately 6.0. Both sets of data are compatible with the hypothesis that four lysine residues are incorporated into one desmosine or isodesmosine moiety. The authors also showed that other labeled amino acids such as [^{14}C]aspartic acid, [^{14}C]glutamic acid and [^{14}C]threonine were not incorporated into desmosine or isodesmosine. In the latter paper, Anwar and Oda[74] employed [1-^{14}C]lysine and [6-^{14}C]lysine in place of the uniformly labeled amino acid. Again the data obtained from the alkaline degradation studies and the ninhydrin–CO_2 studies were entirely compatible with the incorporation of four lysine residues into one desmosine or isodesmosine. Such findings strongly support the hypothesis that the carboxyl groups of the lysine residues involved in the synthesis are maintained intact in the resulting desmosines. That is to say, formation of desmosine and isodesmosine must occur at the opposite end of the lysine residues (carbon 5 and 6).

Sandberg and Cleary[75] studied the changes with time in the specific activities of radioactive components of aortic elastin of young, rapidly growing chicks following a single intravenous injection of uniformly labeled lysine. 10-day-old chicks received single injections of 1000 μC [$^{14}C_6$]lysine per animal. The animals were sacrificed over a period of 40 days. Elastin prepared from the aortas of these animals were analyzed for quarter-desmosines and lysine. Appreciable radioactivity in the desmosines was found after 1 h. The product–precursor relationship of lysine and desmosine was examined using the weighting function technique of Beck and Rescigno[76]. It was proposed within the limitations imposed by frequency of sampling, that lysine to desmosine transformation involves a single intermediate compound. The authors also questioned the validity of the Zilversmit rule[77] in treatment of the data of Partridge et al.[71] and indicated that although it seemed likely from the structural formula, it was not proved that desmosine and isodesmosine were derived from four lysines. It does not appear that the authors were aware of the work of Anwar and Oda mentioned above. One must conclude that four lysine residues previously incorporated into the peptide backbone are incorporated into one desmosine or isodesmosine moiety.

That lysine is also the precursor of lysinonorleucine was proposed by Franzblau et al.[66]. In studies extending those of Miller et al.[70] on developing chick embryos, the concentration of lysinonorleucine in aortic elastin was found to increase at the same rate as that of the desmosines. Results obtained by using 12-day-old chick embryo aortas in tissue culture substantiated the hypothesis that two lysine residues are incorporated into one lysinonorleucine molecule. Briefly, [^{14}C]lysine was added to the tissue culture medium

TABLE IV

SPECIFIC ACTIVITIES OF SEVERAL AMINO ACIDS OBTAINED FROM [^{14}C]LYSINE LABELED ELASTIN

Amino acid	counts/min per μmole[a]	
	Not reduced	Reduced
Lysine	34000	31000
Desmosine	27000	} 44000
Isodesmosine	31000	
Lysinonorleucine	33000	37000
Merodesmosine	—	31000
Lysinoalanine	22000	18000

[a] Based on lysine equivalents assuming: (a) desmosine and isodesmosine equal to four lysine equivalents; (b) merodesmosine equal to three lysine equivalents; and (c) lysinonorleucine equal to two lysine equivalents. See text for explanation.

for 1 day and chased with nonradioactive medium for an additional 8 days. During this time the medium was changed every other day. Elastin was isolated from the organ culture and hydrolyzed, and the hydrolysate was placed on an amino acid analyzer equipped with a stream-splitting device. The portion of the eluant not analyzed with ninhydrin reagent was counted in a liquid-scintillation device. The specific activities obtained for lysine, the desmosines, and lysinonorleucine are given in Table IV. The data confirm the results of other laboratories that four lysine molecules are incorporated into the desmosines. It is also evident from these data that two lysine molecules are condensed to form one lysinonorleucine.

The data of Sandberg and Cleary[75] previously mentioned showed that radioactive lysine is incorporated into the lysinonorleucine of elastin obtained from growing chicks.

(d) Mechanism of crosslink formation

Knowing that lysine is the precursor of these special amino acids, one may then ask how they are synthesized. The cell or cell type responsible for the biosynthesis of elastin is still obscure. At present there appear to be two schools of thought — one implicates muscle cells and the other fibroblasts (see the section on Biogenesis for a more complete discussion, p. 699). In either case the first step in the biosynthesis of elastin would require formation of a soluble "pro-elastin", which should be recognizable by its amino acid composition. Compared with mature elastin, a "pro-elastin" should contain significantly less desmosine, isodesmosine, and lysinonorleucine, but relatively more lysine. The "pro-elastin" could be incorporated into an insoluble elastic fiber via formation of a desmosine or isodesmosine crosslink through the following series of events: (1) An enzymatic deamination of the ε-amino groups of specific lysine residues along the polypeptide chain leads to formation of residues of α-aminoadipic acid-δ-semialdehyde. (2) Three of the aldehyde residues, arising either on different chains or on the same chain, brought into juxtaposition with a fourth intact lysine residue, condense to form the carbon and nitrogen skeleton of the desmosines. (3) If any of the three aldehyde residues implicated or the intact lysine residue is present in the insoluble fiber initially, the soluble "pro-elastin" would then by this process be incorporated into the insoluble elastic network. The possibility is not precluded that additional desmosine crosslinks can be formed in the insoluble elastin fiber by the same mechanism. Similar schemes have been

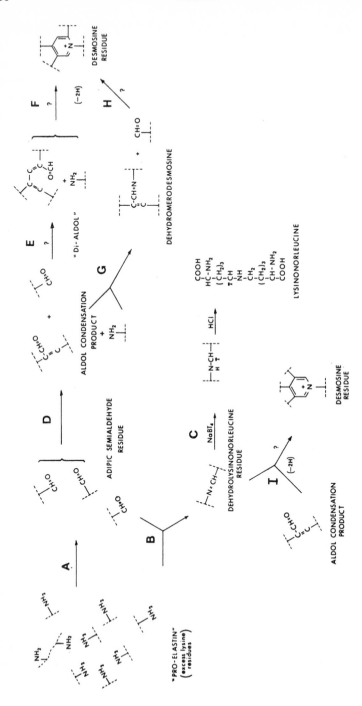

Fig. 5. Possible interrelated scheme for biosynthesis of the crosslinks of elastin. See text for details.

presented by Partridge[78], Piez[6] and Anwar and Oda[72]. Such a scheme is outlined in Fig. 5.

Franzblau et al.[67] proposed that the synthesis of lysinonorleucine could occur with certain similarities. The proposed biosynthetic scheme would include: (1) enzymatic deamination of the ε-amino groups of one residue of lysine to yield a residue of α-aminoadipic acid-δ-semialdehyde, (2) condensation of the aldehyde residue with the ε-amino group of a second residue of lysine to form a Schiff base (i.e. a residue of $\Delta^{6,7}$-dehydrolysinonorleucine) resulting in crosslinking of two elastin chains or two segments of a single chain; and (3) reduction of the aldimine function of the Schiff base to produce the final crosslinking residue of lysinonorleucine. This scheme is also included in Fig. 5.

(e) Intermediates in crosslink formation

In support of the above scheme for the biosynthesis of lysinonorleucine Franzblau and Lent[79] have obtained evidence that the proposed Schiff-base intermediate is present in elastin and can be reduced chemically to lysinonorleucine according to the reactions shown[79] in Fig. 6. Since sodium borohydride can reduce carbonyl compounds to alcohols, Schiff bases to secondary amines, and pyridinium compounds to dihydropyridine compounds[80],

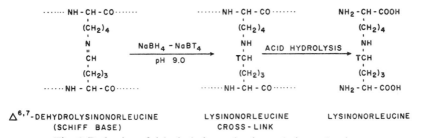

Fig. 6. Reduction of dehydrolysinonorleucine to lysinonorleucine.

the use of this reagent would lead to formation of derivatives more stable than the parent intermediates. The use of tritiated sodium borohydride ($NaBT_4$), as described by Blumenfeld and Gallop[81], has the additional advantages of providing a means of quantitation of extent of reduction and of adding a radioactive marker to the components in question. Thus, treatment of elastin with NaBT$_4$ followed by hydrolysis resulted in the introduction of tritium into fractions, separable on the column of the amino acid analyzer, containing lysinonorleucine. Indeed an increase of total content of lysinonorleucine was observed as compared with the amount in a control (nonreduced) elastin, which specifically demonstrated the initial presence of the Schiff-base

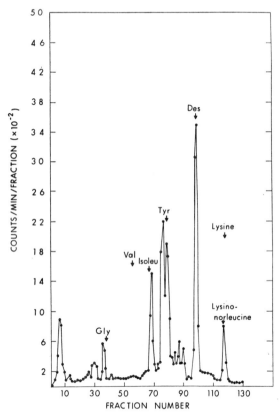

Fig. 7. Distribution of tritium from reduced bovine elastin (NaBT₄) hydrolyzed in 6 N HCl. Arrows indicate the positions of certain amino acids eluting from the amino acid analyzer.

precursor. Tritium also was incorporated into the desmosine fractions and into certain other compounds discussed below, which suggested that the desmosines and these other components of elastin were reduced under the conditions used. A typical chromatogram obtained from elastin following these procedures is shown in Fig. 7. All the peaks arising from this treatment were subsequently shown to originate from lysine residues previously incorporated into the elastin by tissue culture procedures[82].

Evidence that the deamination of certain lysyl residues of elastin and collagen to residues of α-aminoadipic acid-δ-semialdehyde residues is enzymatically catalyzed has been presented by Pinnell and Martin[83]. This will be discussed in more detail below.

Concerning the biosynthetic mechanism for the formation of desmosine and isodesmosine, it should be obvious from the scheme presented in Fig. 5 that several aldehyde intermediates may be involved. Miller and Fullmer[16], using an aldehyde method first described by Sawicki et al.[84] and modified by Paz et al.[85] for proteins, showed that elastin from chick aortas obtained by mild purification procedures contained aldehydes. The method, which involves the use of N-methylbenzothiazolone hydrazone, yielded a spectrum from the solubilized treated elastin which indicated the presence of both saturated and unsaturated aldehydes. In addition these authors reacted 2,4-dinitrophenylhydrazine with elastin. Both methods indicated that there were approximately 6 aldehyde residues per 1000 amino acid residues in aortic elastin from chick. The use of model compounds suggested the ratio of saturated aldehyde to unsaturated aldehyde to be approximately 2.0.

TABLE V

ANALYSES OF BOVINE ELASTIN AND ITS 2,4-DINITROPHENYLHYDRAZONE DERIVATIVE BEFORE AND AFTER REDUCTION[a]

Treatment	Residues/1000 residues amino acid		Total counts recovered		
	Lysinonor-leucine[b]	2,4 DNP hydrazone[c]	Complete analysis	Desmosine fractions	Lysinonorleucine fractions
None	0.85	0	0	0	0
Reduction (NaBT₄)	1.15	0	160000	35000	7700
2,4-DNPH	0.80	10.2	0	0	0
Reduction (cold NaBH₄) followed by 2,4-DNPH	1.10	0.15	—	—	—
2,4-DNPH followed by reduction (NaBT₄)	0.85	10.2	52000	40600	1050

All reductions were carried out with the same batch of standard sodium borohydride mixture
The ninhydrin color equivalent of lysinonorleucine is twice that of leucine.
Based on Kjeldahl nitrogen assuming an average residue weight of 100.

Franzblau and Lent[79] showed that if bovine elastin was treated first with 2,4-dinitrophenylhydrazine (2,4-DNPH), subsequent treatment with $NaBT_4$ (described above) did not cause incorporation of tritium into lysinonorleucine; in fact only the desmosines were reduced and labeled. Conversely, if reduced elastin was subsequently treated with 2,4-DNPH, little or no hydrazone formation occurred. The results are summarized in Table V. Reduction caused about 25% increase in the content of lysinonorleucine determined chromatographically. The magnitude of this increase is highly significant because the extent of radioactivity incorporated into the lysinonorleucine also indicated a 25% increase in lysinonorleucine content. The number of aldehydes found in elastin from *ligamentum nuchae* was found to be higher than the results of Miller and Fullmer[16], and Piez[6] commented that both sets of data were probably high due to interfering substances. As will be seen below, this does not appear to be the case.

Reaction of 2,4-DNPH with elastin before reduction should produce DNP-hydrazones of carbonyl and Schiff-base components present in the protein if these are accessible to the reagent. Treatment of the DNP-hydrazone derivative of elastin with sodium borohydride, in contrast to elastin *per se*, would not result in reductive formation of alcohols and secondary amines since the precursor aldehydes and Schiff bases, respectively, would be present as DNP derivatives.

These results support the conclusion that carbonyl and Schiff-base components are present in the bovine elastin preparation utilized. Reduction of desmosine or isodesmosine must occur in the pyridinium portion of the molecule, since treatment of elastin with 2,4-DNPH had no apparent effect on the extent of this reduction. Incorporation of tritium in the desmosine fractions is accompanied by a skewing of chromatographic peaks obtained subsequently, so that discrete desmosine and isodesmosine peaks are no longer observed. Also, absorption of light at 275 mμ decreases significantly compared with that shown by the desmosines before borohydride treatment.

The N-methylbenzothiazolone reagent has also been utilized as a histochemical stain for both elastin and collagen[16].

Starcher, Partridge and Elsden[86] reported the presence of small amounts of a new amino acid in hydrolysates of bovine elastin which had been alkali treated and then reduced with $NaBH_4$. This compound, which they named merodesmosine, has a structure (shown in Fig. 8) that can be perceived to result from three lysine molecules with the loss of two ε-amino groups. It has been proposed by these authors as a possible intermediate in the desmosine biosynthesis. It should be pointed out that formation of merodesmosine according to Starcher et al.[86] requires pretreatment of the purified elastin with alkali before reduction. If reduction was carried out without the pretreatment of alkali little or no merodesmosine was found. Smith and Franzblau[87] have been able to detect merodesmosine in the developing chick embryo aorta by reduction alone. More merodesmosine is produced when the elastin preparations from chick embryo aorta are pretreated with alkali and then reduced with $NaBH_4$. Starcher et al.[86] proposed that merodesmosine arises

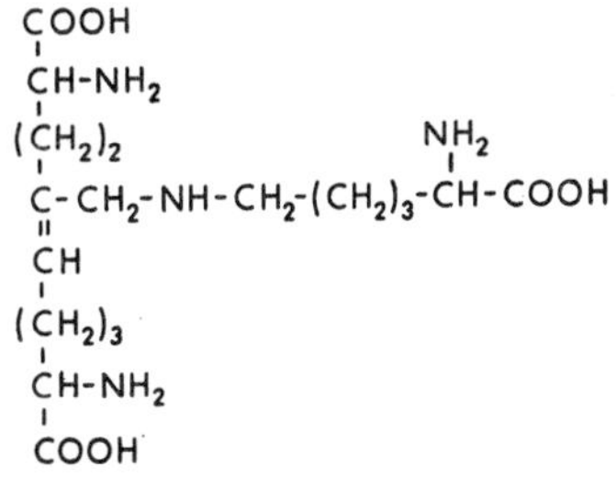

$$
\begin{array}{l}
COOH \\
| \\
CH\text{-}NH_2 \\
| \\
(CH_2)_2 \qquad\qquad NH_2 \\
| \qquad\qquad\qquad\quad | \\
C\text{-}CH_2\text{-}NH\text{-}CH_2\text{-}(CH_2)_3\text{-}CH\text{-}COOH \\
\| \\
CH \\
| \\
(CH_2)_3 \\
| \\
CH\text{-}NH_2 \\
| \\
COOH
\end{array}
$$

Fig. 8. Structure of merodesmosine.

from condensation of the ε-amino group of a lysine residue with the aldol condensation product of two molecules of α-aminoadipic acid-δ-semialdehyde to form a Schiff base which is then reduced by reaction with $NaBH_4$. It is likely that this Schiff base rather than the reduced compound itself would act as an intermediate in the synthesis of desmosine or isodesmosine.

Miller et al.[88] identified one of the aldehydes in elastin as the δ-semialdehyde of α-aminoadipic acid. By oxidation of elastin preparations with performic acid they were able to isolate after hydrolysis in 6 N HCl the corresponding α-aminoadipic acid. The oxidation was carried out according to the procedure of Moore[89] and the resulting α-aminoadipic acid was verified in two separate chromatographic systems as well as on the amino acid analyzer. It is not certain how much of the α-aminoadipic acid would have arisen from the oxidation of certain intact lysyl residues, desmosine and isodesmosine residues as well as other intermediates in the biosynthesis of the desmosines.

If as suggested by these experiments, oxidation of [¹⁴C]elastin with performic acid leads to α-aminoadipic acid then reduction should lead primarily to ε-hydroxynorleucine, since the residues being oxidized or reduced, as the case may be, would be those of α-aminoadipic acid-δ-semialdehyde. Studies by Lent and Franzblau[90] on the reduction of elastin with $NaBT_4$ indicated the formation of three significant radioactive fractions in addition to those of the lysinonorleucine and the desmosines. These are in the region where glycine, isoleucine and tyrosine normally appear on the amino acid chromatograms. Lent et al.[82] addressed themselves to the question of whether these other fractions had their origin in pre-existing lysine residues of the elastin. Elastin which was labeled with [^{14}C]lysine in tissue culture was isolated and reduced with $NaBT_4$. It was found that those radioactive peaks which appear in the region of glycine, leucine and tyrosine contained ^{14}C in addition to tritium when $NaBT_4$ was used. This suggested that these compounds were, in fact, derived from lysine. Since these compounds are not present in a hydrolysate of elastin unless reduction precedes hydrolysis, one may conclude that they exist actually as carbonyl compounds and/or Schiff bases. Acid hydrolysis would almost certainly destroy the integrity of such compounds.

The peak which eluted near glycine was shown to be ε-hydroxynorleucine. However the yields were extremely low. This was also shown by Partridge in his studies[91]. It turns out that ε-hydroxynorleucine is not stable to hydrolysis in 6 N HCl and is converted in 70% yield to ε-chloronorleucine. Lent et al.[82] showed the presence of ε-chloronorleucine in acid hydrolysates of reduced bovine elastin. It elutes in the region near tyrosine and therefore accounted for one of the radioactive peaks appearing in the region where tyrosine is eluted. Treatment of such a hydrolysate with dilute alkali caused the formation of pipecolic acid from the ε-chloronorleucine. These authors found that hydrolysis of reduced elastin in 2 N NaOH resulted in complete stability of the ε-hydroxynorleucine. From such hydrolysates the content of ε-hydroxynorleucine could be quantitated. There were 2–3 residues of ε-hydroxynorleucine per 1000 amino acids in an alkaline hydrolysate of reduced elastin from bovine ligamentum nuchae.

It was noted by Lent et al.[82] that alkaline hydrolysis of reduced bovine elastin not only led to complete recovery of ε-hydroxynorleucine, but the elution pattern of radioactivity on the amino acid analyzer indicated several other striking differences from the elution pattern of an acid hydrolysate (Fig. 9). The radioactive peaks near the tyrosine and isoleucine regions were not present and a new radioactive peak eluted near phenylalanine. This peak contained far more counts than those found in the ε-hydroxynorleucine or

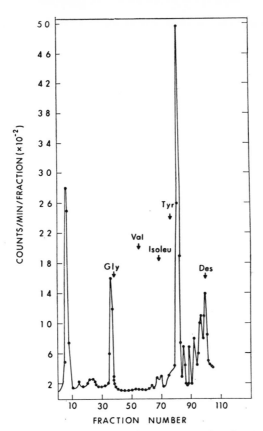

Fig. 9. Distribution of tritium in 2 N NaOH hydrolysate of bovine elastin reduced with NaBT$_4$. Arrows indicate positions of certain amino acids eluting from the amino acid analyzer.

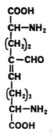

Fig. 10. Structure of aldol condensation product from elastin.

in the desmosines. The purified component was shown to be the reduced aldol condensation product of two residues of α-aminoadipic acid-δ-semialdehyde. The structure of the unreduced aldol condensation product is shown in Fig. 10. The structure was determined by mass spectrometry of the ethyl ester derivative and by a study of the oxidation product of the compound with periodate and permanganate[92]. Such oxidation leads to the formation of aspartic acid, glutamic acid and α-aminoadipic acid as well as the semialdehydes of glutamic and α-aminoadipic acids. Amino acid analyses indicated that 4–5 residues of this aldol condensation product are present per 1000 amino acid residues. Since the aldol is present in such large quantities (there are only 3 desmosine plus isodesmosine residues), the authors suggested that it may serve as an independent crosslink in elastin as well as an intermediate in the biosynthetic pathway of the desmosines. That radioactive lysine is incorporated into the aldol condensation product in tissue culture was demonstrated by Salcedo *et al.*[93].

One concludes from these data that several possible intermediates may be involved in the biosynthetic pathway to desmosine and isodesmosine. In Fig. 5 all the possible intermediates as well as the final crosslinking products are included.

Step (A). It is generally agreed that the initial step for crosslink formation involves the deamination of certain lysyl residues to α-aminoadipic acid-δ-semialdehyde residues.

Step (B). If one lysyl residue and one α-aminoadipic acid-δ-semialdehyde residue combine to form a Schiff base, subsequent reduction,

Step (C), would lead to formation of lysinonorleucine.

Step (D). The formation of desmosine or isodesmosine could conceivably arise from three different pathways all of which involve the formation of aldol condensation product of two residues of α-aminoadipic acid-δ-semialdehyde.

Step (E). The aldol condensation product combines with another residue of α-aminoadipic acid-δ-semialdehyde to form a "dialdol condensation product" which in turn,

Step (F), combines with an intact lysyl residue to form the carbon and nitrogen skeleton of desmosine or isodesmosine.

Step (G). Alternatively, the aldol condensation product combines with an intact lysyl residue to form a dehydromerodesmosine which then,

Step (H), combines with another α-aminoadipic acid-δ-semialdehyde residue to form the carbon and nitrogen skeleton of the desmosines.

Step (1). A third possibility involves the direct condensation of the Schiff base form of lysinonorleucine and the aldol condensation product to form the carbon and nitrogen skeleton of desmosine and isodesmosine.

Piez[6] first pointed out that any of the pathways leading to desmosine or isodesmosine formation must involve in addition to the three deaminations of lysines the removal of two hydrogen atoms from the condensation product mentioned above. This is illustrated in Fig. 5.

9. Experimental disorders of connective tissues as aids to investigation of the crosslinks of elastin

(a) Lathyrism

The pathological effects of feeding aminonitriles such as β-amino-propionitrile (BAPN) to several laboratory animals including rats, chicks and guinea pigs have been known for many years[94]. Osteolathyrism, which is the name given to this experimental disease, is characterized by certain alterations in the integrity of connective tissue protein. The connective tissue appears to lose much of its strength and stability and frequently death is encountered in the animal *via* a dissecting aneurysm. The role of elastin in this connective tissue disease was usually not considered a primary target for study, while studies on collagen impairment were almost always carried out. As the chemistry and biosynthesis of elastin crosslinks was unfolding little was known of the chemistry of crosslinks in collagen, although it had been postulated by several workers that BAPN did interfere with the crosslinking of collagen. Miller *et al.*[95] included the use of BAPN in their initial tissue culture studies which involved the incorporation of [^{14}C]lysine in the desmosine and isodesmosine moieties. It was found that in the presence of the lathyrogen little or no ^{14}C-labeled desmosine was formed and relative to those cultures without BAPN there was no significant increase or decrease in the content of uniformly labeled [^{14}C]lysine in the isolated elastin hydrolysates. O'Dell *et al.*[96] found similar results with aortas from ducklings 1–7 days old grown in organ culture. Miller and Fullmer found that in chicks fed BAPN the isolated aortic elastin contained a higher content of lysine residues with no significant change in the concentration of the desmosines and a decrease in the aldehyde content of the lathyritic elastin when compared to normal control animals. Studies in the author's laboratory have revealed that in the presence of BAPN no lysinonorleucine is synthesized in tissue culture as well. Salcedo *et al.*[93] have also shown that the presence of BAPN in tissue culture inhibits the formation of the aldol condensation product of two residues of α-aminoadipic acid-δ-semialdehyde. No additional residues of α-aminoadipic acid-δ-semialdehyde residues were produced either. These data are compatible with the hypothesis that BAPN affects the first step in the synthesis of elastin crosslinks, namely, the deamination of lysyl residues to α-aminoadipic acid-δ-semialdehyde residues. These data explained for the first time the molecular consequences of feeding BAPN to a laboratory animal. It soon became apparent that similarity in the crosslinking mechanism of collagen and elastin did exist and that the action of BAPN was indeed the same with both of these connective tissue proteins. The differences and similarities in crosslinks will be discussed in detail below.

To describe the mechanism by which BAPN actually inhibits the formation of α-aminoadipic acid-δ-semialdehyde residues from lysyl residues, two possibilities have been proposed. The first really involved studies on collagen. Levene[97] suggested that the BAPN combines with certain aldehyde residues in collagen thus preventing them from forming necessary crosslinks. No BAPN or other lathyrogenic agent has ever been found bound to the collagen however. The most recent work of Pinnell and Martin[83] appears to confirm the second general mechanism, namely, that BAPN inhibits the enzyme system responsible for this deamination. Pinnell and Martin[83] have now been able to isolate from extracts of chick embryonic bone a crude enzyme capable of deaminating lysyl residues in connective tissue proteins. The substrate which they employ for assays is essentially a portion of chick-embryo aorta which is grown in tissue culture in the presence of [14C]lysine and BAPN. This yields a substrate which incorporated [14C]lysine, however the lysyl residues are not utilized to form crosslinks in the presence of BAPN. After incubation with the enzyme system described by Pinnell and Martin[83], the substrate is oxidized with performic acid and the hydrolysates are examined for [14C]aminoadipic acid. Alternatively, [6-3H]lysine is utilized in the tissue culture system and the assay involves the liberation of tritiated water in the presence of enzyme. This is illustrated in Fig. 11. It was also shown that BAPN inhibits this enzyme system; in fact, the BAPN appears to be bound to the enzyme. It therefore appears that the mechanism of action of BAPN involves the inhibition of a specific enzyme system capable of deaminating certain lysyl residues in collagen and elastin. As pointed out by the authors the substrate was crude and appeared to contain in addition to other substances both collagen and elastin. Lysine, itself, was not a substrate for this enzyme system. Page and Benditt[98] have studied the effect of BAPN on an amine oxidase from pig serum. They found that BAPN is not only a competitive inhibitor of the enzyme, when tested against kynuramine, but that the BAPN was a substrate by itself.

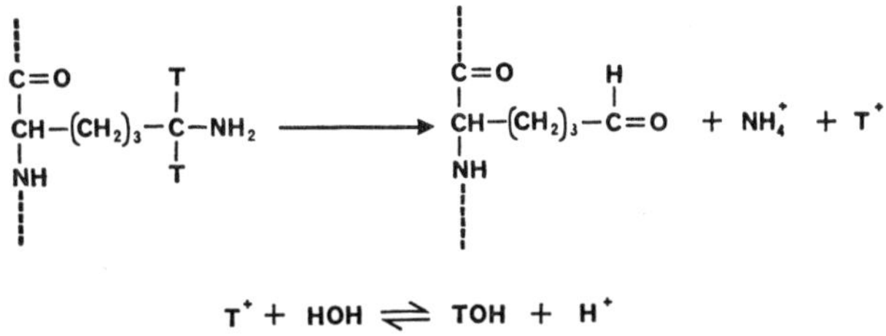

$$T^+ + HOH \rightleftharpoons TOH + H^+$$

Fig. 11. The liberation of tritiated water as catalyzed by lysyl oxidase (Pinnell and Martin[83]).

(b) Copper deficiency

(i) Effect on crosslinking of elastin

The effect of copper deficiency on connective tissue has been known for several years. O'Dell et al.[99] reported that chicks fed copper-deficient diets exhibited both subcutaneous and internal hemorrhage. Aortas of those animals that were autopsied indicated a thickened wall and dissecting aneurysm was usually evident. Elastic fibers within these vessels were disrupted which suggested to these investigators that a major defect in copper deficiency was in the elastic tissue. Studies on copper-deficient swine[100] yielded similar results. All piglets placed on a copper-deficient diet indicated that there were minute breaks in the internal elastic membrane. These changes were observed in all the muscular arteries examined. Starcher, Hill and Matrone[101] studied the effect of copper deficiency on the amino acid content of the aortic elastin from chicks. It was found that the elastin of copper-deficient animals contained three times as much lysine when compared to the control group. It was noted by the authors that Weissman et al.[100] had found similar results in the copper-deficient swine. It was also noted that the elastin from copper-deficient animals solubilized more quickly in formic acid than did elastin from control chicks. It must be pointed out that these workers published their data before the discovery of desmosine and isodesmosine had been published. One clearly can see that the effect of copper deficiency in these studies focuses again on the biosynthesis of the crosslinks. Studies on the lysine content and desmosine content of animals fed copper-deficient diets have since been made and they bear this out. It again turns out that the effect appears to focus on the first step and conversion of lysine to α-aminoadipic acid-δ-semialdehyde[83]. Miller et al.[95] examined the elastin from aortas of chicks fed a copper-deficient diet and essentially confirmed the data of Starcher et al.[101] that there is significantly more lysine in the diseased elastin. In addition they found a slight decrease in the content of desmosine and isodesmosine.

(ii) Isolation of soluble elastin precursor

The isolation of a soluble precursor of insoluble elastin has been attempted in many laboratories. In 1963, Weissman et al.[100] reported on the presence of such a molecule in copper-deficient pigs. Smith et al.[102] reported that this soluble elastin undergoes reversible coacervation, much the same as the α-elastin first described by Partridge et al.[14]. It also was shown to lack the desmosine crosslinks and had a high content of lysine.

Most recently, Sandberg et al.[103] have isolated this soluble precursor of elastin in relatively pure form. Briefly, a freshly removed thoracic aorta from copper-deficient swine 90–120 days old was stripped of its adventitia, minced, and frozen with liquid nitrogen to facilitate crushing the tissue into a fine powder. The tissue was then treated to further homogenization in the presence of two volumes of cold 0.02 M formic acid at pH 2.8. Extraction in the formic acid was carried out for two 24-h periods at 4° with gentle agitation. After concentrating the extracted material, reversible heat precipitation (coacervation) was carried out by allowing the protein solution to warm slowly in a water bath at 25°. The coacervate was removed by centrifugation and the material was further purified by cation-exchange chromatography.

Its amino acid composition resembled insoluble elastin purified from porcine aorta. However, it contained no desmosine or isodesmosine and had a considerably higher content of lysine than in the control. The amino acid composition of the tropoelastin versus the insoluble elastin is given in Table VI. The authors estimated the molecular weight to be 67000 from disc-gel electrophoresis. A thorough investigation of this molecule will be an absolute necessity before one can fully understand the formation of elastin fibers.

It is of interest to note that Petruska and Sandberg[104] were able to predict from previously published amino acid analyses what the composition of soluble and insoluble elastin might be. By assuming that there was no hydroxyproline in elastin they were able to adjust various amino acid analyses

TABLE VI

AMINO ACID COMPOSITION OF PORCINE ELASTIN AND TROPOELASTIN[a]

Amino acid	Porcine tropoelastin	Porcine aortic elastin
Hydroxyproline	11.2	14.5
Aspartic acid	2.9	8.8
Threonine	13.8	7.4
Serine	9.4	8.1
Glutamic acid	18.3	20.9
Proline	108.7	93.8
Glycine	333.7	328.9
Alanine	218.1	233.3
Cystine/2	0.0	0.0
Valine	120.6	124.9
Methionine	0.0	1.7
Isoleucine	18.8	19.6
Leucine	46.2	57.4
Tyrosine	15.6	16.9
Phenylalanine	27.9	32.3
Isodesmosine/4	0.0	8.1
Desmosine/4	0.0	8.3
Lysine	47.5	7.6
Histidine	0.0	1.1
Arginine	7.1	6.6

[a] Taken from Sandberg et al.[103]. Data expressed in residues per 1000 residues.

of elastin to yield a surprising value of 46 lysine residues in the soluble elastin. This value is not far, however, from those reported by Smith et al.[102] and small corrections in collagen content would not lead to large changes in the lysine content. The value obtained by Sandberg et al.[103] was 47 residues of lysine per 1000 amino acids in elastin obtained from the copper-deficient animals.

(iii) *Conversion of lysyl residues in soluble elastin to insoluble elastin fiber*
Knowing what we do now know of the nature and content of the cross-linkages in elastin, it is tempting to see how the data of Sandberg et al.[103] and Smith et al.[102] fit the existing data on the crosslinkages of elastin. The tropoelastin isolated from copper-deficient pigs has a lysine content of 47 residues per 1000 amino acid residues. If this value is meaningful, it should be

possible to account for these lysine residues in the mature elastin fiber. If, at this writing, one can overlook species and tissue differences in the elastin, a reasonable accounting can be obtained. It is necessary to point out certain differences. The content of desmosine and isodesmosine from a variety of sources has been reported to be approximately 3 residues (desmosine plus isodesmosine) per 1000 amino acid residues. Sandberg *et al.*[103] report for the normal control pig a value of approximately 4 residues of desmosine and isodesmosine per 1000 amino acids.

Assuming that one desmosine or isodesmosine is derived from four lysines and one aldol condensate or one lysinonorleucine from two lysine residues one can account for 22–24 lysine equivalents per 1000 amino acids[82]. This value is derived from data obtained from elastin of the nuchal ligament of the ox. In the latter preparation of elastin, the respective values for the desmosines, aldol condensation product and lysinonorleucine are 3, 4–5, and 1 residues per 1000 residues[79]. In the pig the additional desmosine moiety would account for four additional lysine equivalents. If we assume the values for the aldol condensation product and the lysinonorleucine are comparable in the pig aorta and bovine *ligamentum nuchae*, then we could account for 26–28 residues of lysine equivalents. Added to this, the 2–3 residues of α-aminoadipic acid-δ-semialdehyde and 4 residues of lysine present in bovine elastin account for 32–35 lysine equivalents per 1000 amino acid residues. When three desmosines are present, as in *ligamentum nuchae* elastin, the value decreases to 28–31 lysine equivalents. However, Sandberg *et al.*[103] reported the presence of approximately 8 residues of lysine per 1000 amino acids in aortic elastin from pig. If this value is correct and all other previous assumptions are correct, one can account for 36–39 lysine equivalents in the porcine aortic elastin per 1000 amino acid residues. It is possible that, at most, one additional aldehyde or Schiff-base component is present which derives from 3 lysine equivalents. A summary of these data is given in Table VII.

It is also of interest to examine the number of aldehydes in elastin. Miller and Fullmer[16] estimated approximately 6 aldehyde residues per 1000 amino acid residues as determined by the MBTH method or by reaction with 2,4-dinitrophenylhydrazine. Lent and Franzblau[90] found approximately 10 residues of aldehyde per 1000 amino acids in bovine *ligamentum nuchae*. These values are consistent with the presence of 2–3 α-aminoadipic acid-δ-semialdehyde residues and 4–5 aldol residues per 1000 amino acid residues. This would yield a total of 6–8 aldehydes per 1000 amino acids.

TABLE VII

SUMMARY OF LYSINE-DERIVED AMINO ACIDS IN PORCINE ELASTIN

Amino acid	Res./1000 res.	Lys. equiv./1000 res.
Lysine[a]	8	8
Aminoadipic semialdehyde[b]	2–3	2–3
Aldol condensate[b]	4–5	8–10
Desmosines[a]	4	16
Lysinonorleucine[b]	1	2
Merodesmosine	—	—
Total	19–21	36–39

[a] From Sandberg et al.[103].
[b] From Lent et al.[82]. These data are from bovine *ligamentum nuchae* elastin. See text for details.

(iv) Penicillamine

When penicillamine is injected into chick embryos, as shown by Pinnell et al.[105], there is a striking inhibition of desmosine formation. However, the lysine content of the isolated elastin from penicillamine-treated embryo does decrease in a manner similar to the control embryos. What does show a remarkable increase in the treated embryos is the content of α-aminoadipic acid-δ-semialdehyde. The latter was determined by conversion to adipic acid with performic acid. The authors concluded that penicillamine blocks the biosynthesis of elastin crosslinks at a step beyond the initial deamination of lysine (see Fig. 12).

Nimni and his co-workers[106] had shown previously that penicillamine is capable of interacting in a reversible manner with collagen as well. The possibility of interaction with aldehydes in collagen to form the corresponding thiazolidine was also noted. Franzblau and co-workers[107] have shown that the elastin–penicillamine interaction, *in vitro*, is also reversible. However, if penicillamine treatment is followed by reduction with $NaBT_4$ the interaction is irreversible. That is to say, the penicillamine is bound permanently to the elastin. Examination of the reduced protein after alkaline hydrolysis revealed no ε-hydroxynorleucine. This would be expected if penicillamine had reacted with α-aminoadipic acid-δ-semialdehyde residues. This interaction would interfere with the formation of ε-hydroxynorleucine. The postulated interaction of penicillamine with elastin is illustrated in Fig. 13. The proposed reduction with $NaBT_4$ is also included. Evidence for the formation of such a reduced derivative awaits further investigation.

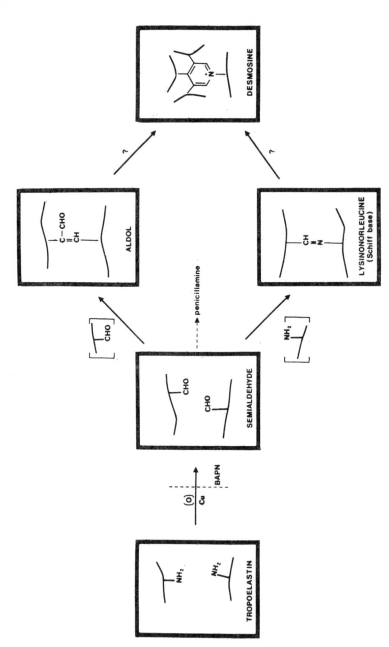

Fig. 12. A schematic illustration of possible metabolic inhibitors of crosslinking in elastin.

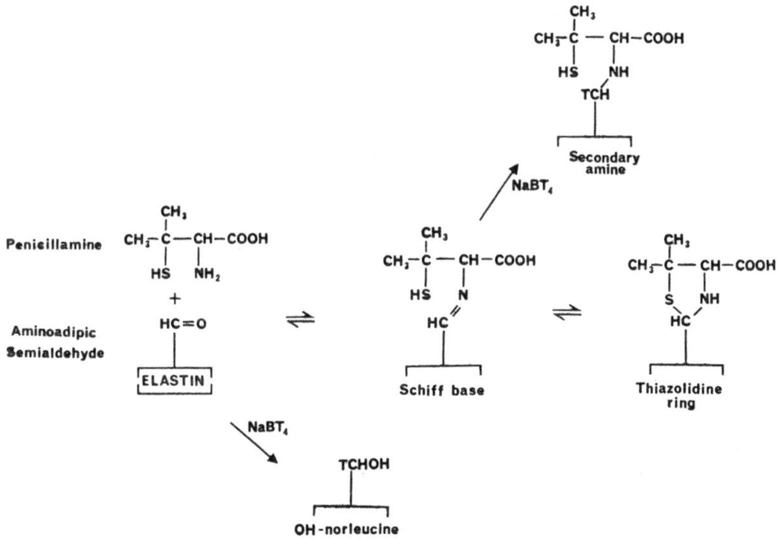

Fig. 13. Possible interaction of penicillamine with elastin.

10. Relationship of crosslinks in collagen to crosslinks in elastin

It is always tempting to workers in the field of connective tissue proteins to compare the properties of elastin to those described for collagen. The presence of hydroxyproline in preparations of elastin has always been noted. The fact that one-third of the amino acid residues in elastin and collagen are glycine is intriguing. Nevertheless, these two proteins are not at all similar in their mechanical properties, nor for that matter do they have similar physicochemical properties. The reader is referred to the review by Bailey[108] for a complete description of collagen.

What about the chemistry of the crosslinks in collagen? It appears that similar, if not identical, crosslinking compounds occur in collagen and elastin. Bailey and Peach[109] have recently isolated ε-hydroxylysinonorleucine from reduced preparations of rat-tail tendon collagen. Its structure (Fig.14) suggests that it was formed *via* a Schiff base between a residue of α-aminoadipic acid-δ-semialdehyde and a hydroxylysine residue. This differs from the formation of lysinonorleucine in elastin since the latter utilizes a residue of lysine instead of hydroxylysine. Tanzer and Mechanic[110] and Kang et al.[111] have now been able to identify small amounts of lysinonorleucine in

References p. 709

preparations of reduced calf skin collagen and reconstituted chick-skin collagen, respectively. Bailey and Peach[109] suggested that ε-hydroxylysino-norleucine functions as an intermolecular crosslink. That is to say, that such a crosslink occurs between tropocollagen units in a collagen fibril. Neither ε-hydroxylysinonorleucine nor lysinonorleucine is present unless reduction is first carried out. This is not true of lysinonorleucine in elastin. In the latter, only a small portion of the lysinonorleucine is in the Schiff-base form, whereas in collagen both the ε-hydroxylysinonorleucine and the lysino-norleucine are entirely in the Schiff-base form.

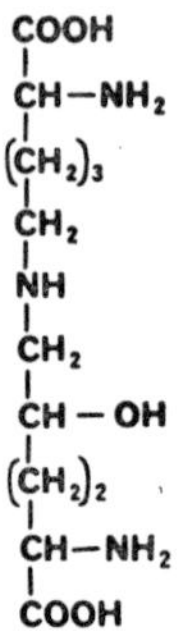

Fig. 14. Structure of hydroxylysinonor-
leucine.

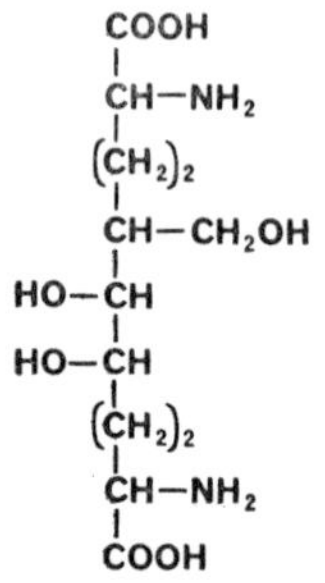

Fig. 15. Structure of reduced aldol condensation product from collagen.

Studies on chick bone collagen and bovine dentine by Bailey et al.[112] revealed still another crosslink after reduction, which was characterized as the aldol condensation product of one residue each of α-aminoadipic acid-δ-semialdehyde and α-aminoadipic acid-δ-hydroxy-δ-semialdehyde (Fig. 15). The latter compound arises from deamination of a residue of hydroxylysine. This again differs from the aldol condensation product previously described in elastin[82] since one hydroxylysine and one lysine residue are involved in the collagen crosslink while two lysyl residues are in the elastin crosslink. Bailey et al.[112] also suggested that this serves as an intermolecular crosslink in collagen.

An intramolecular crosslink in collagen was described by Kang, Faris and Franzblau[113]. The compound which participates in crosslinking two α-chains (α_1 and α_2) to form a β-chain (B_{12}) was shown to be identical with the aldol condensation product found in elastin. Preliminary evidence suggests that this intramolecular crosslink undergoes further condensation during the formation of insoluble elastin[111].

Although the chemical reactivity and physical properties of elastin are so different from collagen their crosslink chemistry is strikingly similar. All of the crosslinks in elastin are derived from lysyl residues and all of the crosslinks described in collagen are derived from a combination of lysyl and hydroxylysyl residues. Since hydroxylysine itself is derived from lysine after incorporation into the peptide chain, one can really say that crosslinking in collagen and elastin derives from lysyl residues incorporated into the backbone polypeptide chains. A list of these crosslinks and their precursors are given in Table VIII.

TABLE VIII

CROSSLINKS IN COLLAGEN AND ELASTIN

Protein	Crosslink	Precursor (residue)
Elastin	Aldol condensate	Two adipic acid-δ-semialdehyde
	Lysinonorleucine	One adipic acid-δ-semialdehyde and one lysine
	Desmosines	Three adipic acid-δ-semialdehyde and one lysine
Collagen	Aldol condensate	One adipic acid-δ-semialdehyde and one hydroxyadipic acid-δ-semialdehyde
	Hydroxylysinonorleucine	One adipic acid-δ-semialdehyde and one hydroxylysine (or *vice versa*)
	Lysinonorleucine	Same as in elastin
	Others	? ?

11. Biogenesis

The fibroblast is generally accepted as being responsible for the production of elastin and in the case of vascular elastin, the smooth muscle cell. Only the latter type of cell has been shown by electron microscopy to exist in the *tunica media* of normal arteries examined by a number of investigators[114-120], where collagenous and elastic fibers are found in the extracellular material. In the *adventitia* where filaments are found associated with small pieces of elastin, the fibroblasts are believed to be the cell type producing elastin.

There is rapid formation of elastin in the embryo[70,114], in the repair of vascular lesions[121], in the uterus during normal pregnancy and postpartum involution[122] and in healing wounds[123].

The examination of the ultrastructure of aortas of growing animals has provided much of the knowledge regarding the biogenesis of elastin. Electron microscope studies of the aortas of young and aging mice[124], the newborn[119] and adult rat[117], the developing chick embryo[125,126], rabbit[127] and man[127,128] and the *ligamentum nuchae* of calf fetus[129], goat and sheep[127] have given clues as to the origin of elastin.

The histogenesis of the developing chick embryo aorta has also been thoroughly investigated with the light microscope[130] and the gradual development of elastin and of periodic acid–Schiff (PAS) positive material has also been studied[131,132].

From his electron micrographs of sections of chick embryo aortas stained with lead hydroxide or phosphotungstic acid, Karrer[114] found elastic fibers in the extracellular material of the *tunica media*. These fibers appear as small (*ca.* 300–1200 mμ), moderately dense structures of roundish or somewhat irregular outlines. These fibers are well developed in the older embryos studied (11 days and over), but in a 7-day-old embryo were rarely seen or were very small (100–200 mμ). Even in the oldest aortas studied (18-day embryo) no continuous elastic lamellae were present. Small granules (*ca.* 130 mμ) are seen at the periphery of elastic fibers. Some of the latter appear homogeneous while the others show a definite filamentous substructure. Thin filaments may also be seen in immediate contact with certain elastic fibers.

Resorcin–fuchsin staining for elastin was negative in the 7-day-old aorta but positive in the 14-day-old. Karrer found the embryo aortas to be negative to the PAS stain although Kinukawa[132] reported the presence of PAS-positive material in elastic fibers and aortas of older embryos (13 days and over).

Cells in the younger embryos contain a greatly developed endoplasmic reticulum and so are classified as fibroblasts. In older embryos, the endoplasmic reticulum is less developed and hence these cells are considered as smooth muscle cells. Karrer believes that the fibroblast is transformed into the smooth muscle cell in the older embryo aortas.

The available electron micrographs suggest that elastin is formed from thin filaments in combination with a material of homogeneous appearance. From his observations, Karrer has outlined three possible origins of elastin: (*a*) the close apposition of certain elastic fibres to the plasma membranes of cells might suggest that elastin can arise from the surrounding surface of the smooth muscle cell; (*b*) the location of other fibers at some distance from cells indicates that elastin fibers may grow independently in the extracellular space; (c) elastin may arise by the direct conversion of extruded units of endoplasmic reticulum into the small elastic fibers. An electron micrograph shows these reticulum units as having the approximate size of the smaller elastic fibers and certain small granules connected with elastic fibers as being comparable in size to the presumed RNA particles of the endoplasmic reticulum.

In the newborn-rat aorta, elastin appears to form mainly at the surfaces of the elastic membranes. A vesicular component stands out as the chief morphological feature of the elastin in the newborn as well as in the adult rat. Vesicles in young elastin were seen to be more variable in size but considerably larger than those in the adult (average diameter, 100 Å). Elastin with larger vesicles (300 Å in diameter) is interpreted to be young, recently formed elastin. At the margins of elastic membranes, areas of very young elastin with large vesicles can be seen to merge gradually with the diffuse flocculent material in the intercellular space. In some places within the elastic membranes of the newborn, patches of young elastin and more mature elastin with smaller vesicles can be seen. Formations as these suggest that the elastin undergoes a maturation process during which the vesicular components become progressively smaller so that at birth, half of the elastin may be relatively mature. In the young adult, almost all the elastin in the membranes is of the mature type. This elastin is PAS-negative in contrast to young elastin which is PAS-positive. On the basis of this staining reaction one can conclude that elastin of the newborn is rich in mucopolysaccharides which may be localized in the large vesicles. This distribution is altered in older elastin. A similar relationship has been reported by Dyrbye and Kirk[133] who found that the total mucopolysaccharide content of young aortas in children was higher than in the aortas of adults.

Electron microscope studies[124] on the thoracic aortas of mice (3 days to 17 months) showed the following age changes: the continuous sheets of elastic laminae which are already essentially complete in young mice (3 days) are gradually separated from the cells by the increasing numbers of collagen fibrils. They become discontinuous, frayed and uneven, and show small lacunae or indentations within the elastin. There is no marked increase in the elastin noticeable during most of the life span, with the exception of the first few days. Most of the elastin is probably laid down during the embryonic age and within a relatively short period after birth.

Earlier, Smith et al.[134] studied the age changes within the mouse aorta

References p. 709

using the light microscope. In mice 2 h to 700 days old, they observed a gradual increase of collagenous fibers and of the ground substance with age, resulting in a thicker aortic wall, whereas the width of the elastic laminae remained constant after the age of 20 days.

Schwarz[135], in his study of elastogenesis in tissue culture, described elastin as first appearing in the shape of droplets with 40–50-Å filaments at the periphery and 80–110-Å filaments within it.

Haust et al.[128] showed from electron microscope studies of fetal, neonatal and adult aortas in man that elastic fibers and lamellae develop by aggregation and fusion of single, characteristic elastin units which show a close relationship to the basement membrane of the smooth muscle cells in *intima* and *media*. Microfibrils found in the extracellular space take part in the organization of elastic units.

The *ligamentum nuchae* of 180-day calf fetus, an age at which intensive elastogenesis begins, was studied by Fahrenbach et al.[129]. They reported that the first ultrastructural indication of elastogenesis consists of small extracellular accumulations of hollow-appearing filaments lying adjacent to fibroblasts and more or less parallel to the long axis of the ligament. These filaments are referred to by these investigators as pre-elastin filaments to distinguish them from elastin fibers which may or may not have a coating of the filaments. Definitive elastin appears within clumps of pre-elastin filaments in the shape of small fibers also oriented in the long axis of the ligament. These elastin fibers measure 700–1000 Å in the 180-day-old fetus ligament and are surrounded by pre-elastin filaments. They form bundles and anastomose with each other thus gradually filling the interstices. These bundles (average diameter, 0.4 μ) represent the early stage of the adult elastin fiber which measures from 3.6 to 9.9 μ in diameter.

Spiny vesicles called acanthosomes have repeatedly been observed associated with developing elastin. The extensive Golgi cisternae of the fibroblasts bud off large numbers of acanthosomes, this activity increasing in the 180-day fetus when adult elastin is beginning to form in significant quantities. Acanthosomes fuse with each other and approach the surface of the cell where they are liberated in the immediate vicinity of developing elastin fibers. After fusing with the cell membrane, the vesicle opens up and flattens out, its filamentous content often in continuity with adjacent elastin. They are also released to a lesser degree near developing collagen and quite frequently between adjacent fibroblasts resulting in adhesion to maintain a close relationship between fibroblasts or elastin and neighboring fibroblasts.

The acanthosomes seen in calf-fetus *ligamentum nuchae* bear a resemblance to the vesicles in rat aorta which were mentioned earlier. The opened acanthosomes at the cell surface are being interpreted as signalling the release of mucopolysaccharides to bind adjacent cells together, to form the ground substance associated with collagen and to enter the process of elastogenesis in a still unknown manner.

Studies on the changes in chemical composition during development of bovine nuchal ligament by Cleary *et al.*[136] showed that insoluble elastin concentration was low until the end of the seventh fetal month, from which time it began to rise rapidly. During the next 10–12 weeks the rate of increase in elastin concentration remained high and reached the adult value. The authors conclude that the immediate postnatal period would be the best time to isolate the soluble precursor elastin.

Studies were done by Greenlee *et al.*[137] on the fine structure of developing elastic fibers in bovine *ligamentum nuchae* (1–9 months fetus) and rat flexor digital tendon (15- and 18-day-old fetus and from newborn, 5- and 30-day-old Sprague weanling rats). In sections stained with uranyl acetate and lead, elastic fibers were found to contain two distinct morphologic components: tubular fibrils of 110-Å diameter and a central, almost amorphous nonstaining area. The first identifiable elastin fibers are composed of aggregates of fine fibrils approximately 100 Å in diameter. With increasing age, the fibrils surround a central amorphous region which seems to stand out more until in mature elastic fibers it is the predominant structure surrounded by the fibrils. The two components have specific staining characteristics: the 100-Å fibrils stain with both uranyl acetate and lead and the amorphous region stains only with phosphotungstic acid. Collagen fibrils stain with uranyl acetate or phosphotungstic acid but not with lead. These staining properties were interpreted to indicate that each structure has a different surface charge, which selectively interacts with each specific stain.

Ross and Bornstein[18] found the central amorphous component to have the same composition and properties as elastin whereas the microfibrils had entirely different properties. This was confirmed by the investigations of Waisman *et al.*[138] on the properties of the microfibrils of the elastic membranes of copper-deficient and normal pig coronary artery. The microfibrils of both are insoluble in water and cold molar sodium chloride and resistant to elastase which completely solubilizes the homogeneous elastin matrix.

More recent studies by both electron microscopy and biochemical means of the elastogenesis of chick embryo aorta by Takagi[125,126] have substantiated the work of Greenlee, Ross and Hartman[137] and the earlier work of Karrer. Sections of aortas of chick embryos 8–20 days old and 3-week chicks were prepared for electron microscopy. Elastogenesis in the aortic media was seen to take place actively in three kinds of medial cells, the undifferentiated mesenchymal cells, fibroblast and smooth muscle cell.

Aortas of 8-day embryos showed only fibroblasts. The extracellular space was narrow and contained small amounts of slender elastic fibers in contact with the cell membranes besides sparse collagen microfibrils in bundles and amorphous materials. Even at this age the two components of the elastic fibers were evident: the white or grey amorphous central core and the fine fibrillar blackish peripheral rims.

The 11-day embryo aortas showed some of the cells abutted by basement membranes as in the smooth muscle cell. Their rough-surfaced cisternae and small vesicles which appeared interconnected by an intricate canal system were frequently situated just beneath the plasma membrane; others were found outside the cells. Newly formed elastic fibers were found in the neighborhood of these vesicles. In other sites, the dilated cisternae contained multilayered electron-dense materials.

The conversion of some medial cells to smooth muscle cells was evident in 15- or 16-day embryos. Their endoplasmic reticulum had regressed leaving a few rough-surfaced cisternae and had well-developed basement membranes. There were some fibroblast-like cells in the media and a few undifferentiated mesenchyme cells underneath the aortic endothelium. The elastic fibers were found in direct contact with the surfaces of the cells, the slender fibers had fused together to form larger ones especially around the smooth muscle cells. The extracellular space gained in width and began to form distinct intercellular layers alternating with the multiplex rings of cell cords. Mucopolysaccharide networks were seen in addition to elastic and collagen fibers.

Elastogenesis seemed to attain its peak on the 17th day of incubation. Abundant masses of elastic fibers were seen around the irregularly shaped cells containing dilated cisternae in the endoplasmic reticulum and next to moderately developed basement membranes. There were large, roundish materials which looked like aggregates of small vesicles circumscribed by plasma membranes and protruding into the extracellular spaces from the tips of cytoplasmic processes. Collagen microfibrils were located at a certain distance from the cells.

Media of 20-day embryos showed a few layers of undifferentiated mesenchymal cells in direct proximity to elastic fibers. Continuous elastic lamellae began to be established. Good amounts of elastic fibers were distributed around smooth muscle cells and fibroblast-like cells.

Throughout the aortic *media* in 3-week chicks, the fibroblast-like cells continued to exist along with the smooth muscle cells, while the undifferentiated mesenchymal cells were still recognizable underneath the endothelium.

These results confirmed an earlier report[139] and are in agreement with the formation of elastic fibers in rat fetus by an extension of the basement membrane around the smooth muscle cell.

Another approach to the study of the biogenesis of elastin was by the examination of the subcellular fractions obtained by differential centrifugation of the aorta homogenate of 15-day chick embryos. Elastin from nuclear fractions appeared homogeneous and contained nearly all of the mature elastic fibers. It consisted mostly of the amorphous, non-staining central core; the fine fibrillar component was observed only in some samples. Its amino acid composition was similar to that generally obtained for elastin.

The elastin from the microsomal fraction was the most homogeneous. Only microsomal vesicles were present. Mature elastic fibers were absent. The isolated elastin showed only clusters of bundle structures which were composed of twisted, fine, tubular fibrils around 100 Å in width. These fibrils are believed to be the contents of the microsomal vesicles. The characteristic amino acid composition of the microsomal elastin is therefore ascribed to this pro-elastin: the larger amount of polar amino acids especially aspartic acid, glutamic acid, lysine and arginine and smaller amounts of the non-polar amino acids as glycine, alanine, valine and proline than in normal elastin, the presence of histidine and absence of hydroxyproline. This is not consistent with the data of Sandberg *et al.*[103].

The mitochondrial fraction contained some microsomal vesicles besides small amounts of the mature elastic fibers. By the same token, the elastin from these fractions consisted of larger amounts of the fibril clusters and small amounts of the amorphous cores; its amino acid composition showed the presence of a mixture of the mature elastin and pro-elastin.

The supernatant fraction yielded an elastin of the same amino acid composition and electron microscope picture as elastin from the microsomal fraction.

Measurement of the uptake of tritiated valine into the various elastin preparations showed that the elastin from the supernatant first attained the highest specific activity, followed by the elastin in the microsomal and mitochondrial fractions. The supernatant fraction elastin was suggested to be a soluble precursor of the pro-elastin.

From what is known at present, elastin may be synthesized by the cell in the form of a soluble precursor which diffuses to the site of fiber formation and there polymerizes to a crosslinked three-dimensional gel.

12. Elastin models

Because of inherent difficulties with purified preparations of elastin in determining physical parameters normally associated with soluble proteins, a well-defined model of elastin is hard to come by. The only kinds of information that one can rely on relate primarily to either the mechanical properties of elastin fibers or to the crosslinking of elastin. If one considers the mechanical properties, which strongly indicate that elastin behaves as a rubber-like elastomer, the structure of elastin, as pointed out so clearly by Partridge[140], can be interpreted in terms of a network of protein chains, crosslinked at intervals, but otherwise free to take up a randomly crumpled configuration giving maximum entropy. Such protein chains carry firmly bound water of hydration and under the influence of thermal motion the hydrated chains are distributed evenly in the space occupied by the liquid water of swelling. Dry elastin is a brittle substance with no rubber-like elasticity. This is usually described as the network model of elastin.

As an alternative to this model, Partridge proposed what is referred to as a corpuscular model[140]. This model incorporates the present knowledge of elastin crosslinks and is generally in keeping with the view of elastin structure arising from electron micrographs of their sections. Kawase[141] reported that such electron micrographs of elastin indicated a bead-like structure along the fiber axis. The fibers appear to lack axial periodicity and measure from 100 Å to 200 Å in width. The beads indicated a diameter equal to their width and the centers of two neighboring beads were separated by a distance of 120 Å to 200 Å. The corpuscular model assumes that insoluble elastin fibers are made from a soluble precursor elastin, *via* the formation of desmosine or isodesmosine crosslinks. Although there are difficulties in imagining just how soluble molecules align properly so that four lysyl residues distributed between the insoluble matrix and the soluble "pre-elastin" are placed in juxtaposition to form a desmosine crosslink, it was suggested that if the proelastin can be thought of as a globular hemoglobin-like structure the specified surface pattern of its tertiary structure may serve to perform this task. As in hemoglobin, hydrophobic groups are directed towards the inside of the globular unit and the hydrophilic groups are directed towards the outside. Further, one can picture that such a molecule contains at its surface pairs of lysine residues which are brought into apposition with those on neighboring molecules at the points of contact. From there those mechanisms previously described for crosslink formation take over. Partridge suggested that a molecular weight of 80 000 for such a molecular model would not be unreasonable. In fact, Sandberg et al.[103] have determined the molecular weight of tropoelastin to be 67 000.

To simplify the hypothesis it was further suggested that each globular precursor molecule contains two independent peptide chains which could have the same or different amino acid composition. The dimer molecule bears on its surface four pairs of lysine residues capable of interacting with other doublets on adjacent molecules to form a desmosine crosslink. When the entire net is completed by crosslink formation a system of corpuscles bound to each other at four points only, is produced.

This arrangement gives rise to a specific tetrahedral arrangement in which the void space amounts to 66% of the total volume if the molecules are spherical. Data presented by Partridge[140] indicate that the water content of swollen bovine *ligamentum nuchae* elastin is 63.1% v/v. A similar model for elastin was suggested by Donovan et al.[142] although the units of the network are not pictured as corpuscular.

At this time, no information concerning the physicochemical properties of tropoelastin has come forth. Attempts to gain insight into the structure of the soluble precursor of elastin have been reported. These studies involve the use of oxalic acid-solubilized elastin. Mammi et al.[143] investigated the infrared spectrum, the far-ultraviolet spectrum and the circular dichroism (CD) spectrum of α-elastin prepared according to the procedure of Partridge et al.[14]. The CD spectra of solutions of α-elastin in water–ethanol mixtures show a continuous increase in α-helix content with increasing alcohol content. At 97% ethanol by volume α-elastin exhibits a characteristic right-hand α-helix spectrum with two negative bands at 228 mμ and 208 mμ and a positive band at about 190 mμ. At best, Mammi et al.[143] estimate only 10% α-helix in water solutions of α-elastin. Urry et al.[144] extended the CD observations of Mammi et al. By causing the α-elastin to form a coacervate at 37° the CD spectrum obtained for the mucilaginous precipitate is characteristic of those obtained for highly helical proteins and polypeptides. These authors estimated that such a coacervate indicated the presence of 50%

α-helical content. One has to point out that such studies must be repeated with the native tropoelastin isolated by Sandberg *et al.*[103], since α-elastin, at best, is a degraded product of native insoluble elastin, which contains many carboxyl and amino groups not normally present in native insoluble elastin[145]. Nevertheless, such experimental models are of extreme value and should be encouraged.

13. Interaction of elastin with other substances

It has been suggested that elastin interacts with several substances *in vivo* ranging from small ions such as calcium ions[146,147] to small molecules such as cholesterol esters[148] to relatively large molecules such as glycoprotein and lipoprotein[149-151]. What the nature of these interactions is, is still not known, although efforts at obtaining such information appear to be increasing. These interactions may be concerned primarily with disease states. For example, the interaction of elastin with cholesterol esters has been implicated in the development of atherosclerosis[152]. Much of the early work of these various interactions has been reviewed in detail[3].

ACKNOWLEDGEMENTS

The author would like to thank L. L. Salcedo, B. Faris and J. Letvinchuk for their untiring help in the preparation of this manuscript, and L. L. Salcedo for her contribution to the section on Biogenesis.

REFERENCES

1   A. N. RICHARDS AND W. J. GIES, *Am. J. Physiol.*, 7 (1902) 93.
2   S. M. PARTRIDGE, *Advan. Protein Chem.*, 17 (1962) 227.
3   J. P. AYER, *Intern. Rev. Connect. Tissue Res.*, 2 (1964) 33.
4   S. SEIFTER AND P. M. GALLOP, in HANS NEURATH (Ed.), *The Proteins*, Vol. IV, Academic Press, New York, 1966, p. 153.
5   I. MANDL, *Advan. Enzymol.*, 23 (1961) 163.
6   K. A. PIEZ, *Ann. Rev. Biochem.*, (1968) 547.
7   W. MÜLLER, *Z. ration. Med.*, III, 10 (1861) 173.
8   J. HORBACZEWSKI, *Z. Physiol. Chem.*, 6 (1882) 330.
9   R. H. CHITTENDEN AND A. S. HART, *Z. Biol.*, 25 (1889) 368.
10 J. SCHNEIDER AND A. HÁJEK, *Biochem. Z.*, 195 (1928) 403.
11 G. M. HASS, *A.M.A. Arch. Pathol.*, 34 (1942) 807.
12 O. H. LOWRY, D. R. GILLIGAN AND E. M. KATERSKY, *J. Biol. Chem.*, 139 (1941) 795.
13 A. I. LANSING, T. B. ROSENTHAL, M. ALEX AND E. W. DEMPSEY, *Anat. Record*, 114 (1952) 555.
14 S. M. PARTRIDGE, H. F. DAVIS AND G. S. ADAIR, *Biochem. J.*, 61 (1955) 11, 21.
15 V. D. HOSPELHORN AND M. J. FITZPATRICK, *Biochem. Biophys. Res. Commun.*, 6 (1961) 191.
16 E. J. MILLER AND H. M. FULLMER, *J. Exptl. Med.*, 123 (1966) 1097.
17 F. S. STEVEN AND D. S. JACKSON, *Biochim. Biophys. Acta*, 168 (1968) 334.
18 R. ROSS AND P. BORNSTEIN, *J. Cell Biol.*, 40 (1969) 366.
19 A. C. BURTON, *Physiol. Rev.*, 34 (1954) 619.
20 E. WOHLISCH, W. J. SCHMIDT, H. WEITNAUER, W. GRÜNING AND R. ROHRBACH, *Kolloid Z.*, 104 (1943) 14.
21 K. H. MEYER AND E. FERRI, *Arch. Ges. Physiol.*, 238 (1936) 78.
22 W. T. ASTBURY, *J. Intern. Soc. Leather Trade Chemists*, 24 (1940) 69.
23 C. A. J. HOEVE AND P. J. FLORY, *J. Am. Chem. Soc.*, 80 (1958) 6523.
24 D. P. MUKHERJEE, *Ph. D. Dissertation*, Massachusetts Institute of Technology, 1969.
25 P. J. FLORY, *Principles of Polymer Chemistry*, Cornell University Press, Ithaca, 1953, p. 471.
26 L. ROBERT AND N. POULLAIN, *Arch. Mal. du Coeur et des Vaisseaux, Suppl., Rev. de l'Athéroscler. et des Arterio. Path. Periphér.*, 8 (1966) 121.
27 A. I. LANSING, in C. RAGAN (Ed.), *Second Conference on Connective Tissues*, Josiah Macy Jr. Foundation, New York, 1951, p. 45.
28 E. W. DEMPSEY AND A. I. LANSING, *Intern. Rev. Cytol.*, 3 (1954) 437.
29 D. P. VARADI AND D. A. HALL, *Nature*, 208 (1965) 1224.
30 R. E. NEUMAN AND M. A. LOGAN, *J. Biol. Chem.*, 186 (1950) 549.
31 M. L. R. HARKNESS, R. D. HARKNESS AND D. A. MCDONALD, *Proc. Roy. Soc. (London), Ser. B*, 146 (1957) 541.
32 R. A. GRANT, *Biochem. J.*, 97 (1965) 5C.
33 R. S. BEAR, *Advan. Protein Chem.*, 7 (1952) 69.
34 F. S. LABELLA, *Arch. Biochem. Biophys.*, 93 (1961) 72.
35 A. KÄRKELÄ AND E. KULONEN, *Acta Chem. Scand.*, 13 (1959) 814.
36 F. J. LOOMEIJER, *Nature*, 182 (1958) 182.
37 F. M. SINEX, in N. SHOCK (Ed.), *Biochemical Aspects of Aging*, Columbia Univ. Press, New York, 1962.
38 C. FRANZBLAU, unpublished results.
39 D. J. CANNON, R. PAPAIOANNOU AND C. FRANZBLAU, *Federation Proc.*, 26 (1967) 832.
40 S. KELLER AND I. MANDL, *Biochem. Biophys. Res. Commun.*, 35 (1969) 687.

41 V. SCARSELLI, *Nature*, 184 (1959) 1563.
42 R. L. ENGLE AND E. W. DEMPSEY, *J. Histochem. Cytochem.*, 2 (1954) 9.
43 S. E. BROLIN AND O. HASSLER, *Acta Soc. Med. Upsalien.*, 66 (1961) 65.
44 D. S. JACKSON AND E. G. CLEARY, *Methods Biochem. Anal.*, 15 (1967) 56.
45 R. A. GRANT, *Brit. J. Exptl. Pathol.*, 47 (1966) 163.
46 R. A. GRANT, *Biochem. J.*, 95 (1965) 53P.
47 J. P. BENTLEY AND A. N. HANSON, *Biochim. Biophys. Acta*, 175 (1969) 339.
48 M. J. BARNES, B. J. CONSTABLE AND E. KODICEK, *Biochem. J.*, 113 (1969) 387.
49 R. A. ANWAR, *Can. J. Biochem.*, 44 (1966) 725.
50 M. LEDVINA AND F. BARTOŠ, *J. Chromatog.*, 31 (1967) 56.
51 J. BALÓ AND I. BANGA, *Nature*, 164 (1949) 491.
52 M. A. NAUGHTON AND F. SANGER, *Biochem. J.*, 78 (1961) 156.
53 A. S. NARAYANAN AND R. A. ANWAR, *Biochem. J.*, 114 (1969) 11.
54 L. VISSER AND E. R. BLOUT, *Federation Proc.*, 28 (1969) 407.
55 E. R. BLOUT, personal communication.
56 N. H. GRANT AND K. C. ROBBINS, *Arch. Biochem. Biophys.*, 66 (1957) 396.
57 N. H. GRANT AND K. C. ROBBINS, *Proc. Soc. Exptl. Biol. Med.*, 90 (1955) 264.
58 F. LAMY, C. P. GRAIG AND S. TAUBER, *J. Biol. Chem.*, 236 (1961) 86.
59 D. S. MIYADA AND A. L. TAPPEL, *Food Res.*, 21 (1956) 217.
60 P. BAUMANN, *Chem. Ind. (London)*, (1959) 1498.
61 K. G. IOFFE AND V. M. SOROKIN, *Biokhimiya*, 19 (1954) 652.
62 D. A. HALL AND J. W. CZERKAWSKI, *Biochem. J.*, 80 (1961) 121, 128.
63 L. ROBERT AND N. POULLAIN, *Bull. Soc. Chim. Biol.*, 45 (1963) 1317.
64 S. M. PARTRIDGE, D. F. ELSDEN AND J. THOMAS, *Nature*, 197 (1963) 1297.
65 J. THOMAS, D. F. ELSDEN AND S. M. PARTRIDGE, *Nature*, 200 (1963) 651.
66 C. FRANZBLAU, F. M. SINEX, B. FARIS AND R. LAMPIDIS, *Biochem. Biophys. Res. Commun.*, 21 (1965) 575.
67 C. FRANZBLAU, B. FARIS AND R. PAPAIOANNOU, *Biochemistry*, 8 (1969) 2833.
68 J. C. SPECK, P. T. ROWLEY AND B. L. HORECKER, *J. Am. Chem. Soc.*, 85 (1963) 1012.
69 S. M. PARTRIDGE, D. F. ELSDEN, J. THOMAS, A. DORFMAN, A. TELSER AND P. L. HO, *Biochem. J.*, 93 (1964) 30C.
70 E. J. MILLER, G. R. MARTIN AND K. A. PIEZ, *Biochem. Biophys. Res. Commun.*, 17 (1964) 248.
71 S. M. PARTRIDGE, D. F. ELSDEN, J. THOMAS, A. DORFMAN, A. TELSER AND P. L. HO, *Nature*, 209 (1966) 399.
72 R. A. ANWAR AND G. ODA, *J. Biol. Chem.*, 241 (1966) 4638.
73 R. A. ANWAR AND G. ODA, *Nature*, 210 (1966) 1254.
74 R. A. ANWAR AND G. ODA, *Biochim. Biophys. Acta*, 133 (1967) 151.
75 L. B. SANDBERG AND E. G. CLEARY, *Biochim. Biophys. Acta*, 154 (1968) 411.
76 J. S. BECK AND A. RESCIGNO, *J. Theoret. Biol.*, 6 (1964) 1.
77 D. B. ZILVERSMIT, C. ENTENMAN AND M. C. FISHLER, *J. Gen. Physiol.*, 26 (1943) 325.
78 S. M. PARTRIDGE, *Gerontologia*, 15 (1969) 85.
79 C. FRANZBLAU AND R. W. LENT, Structure, Function and Evolution in Proteins, *Brookhaven Symp. Biol.*, 1968, Vol. 2, (1969) p. 238.
80 P. S. ANDERSON W. E. KRUEGER AND R. E. LYLE, *Tetrahedron Letters*, (1965) 4011.
81 O. O. BLUMENFELD AND P. M. GALLOP, *Proc. Natl. Acad. Sci. (U.S.)*, 56 (1966) 1260.
82 R. W. LENT, B. SMITH, L. L. SALCEDO, B. FARIS AND C. FRANZBLAU, *Biochemistry*, 8 (1969) 2837.
83 S. R. PINNELL AND G. R. MARTIN, *Proc. Natl. Acad. Sci. (U.S.)*, 61 (1968) 708.
84 E. SAWICKI, T. R. HAUSER, T. W. STANLEY AND W. ELBERT, *Anal. Chem.*, 33 (1961) 93.

85 M. A. Paz, O. O. Blumenfeld, M. Rojkind, E. Henson, C. Furfine and P. M. Gallop, *Arch. Biochem. Biophys.*, 109 (1965) 548.
86 B. C. Starcher, S. M. Partridge and D. F. Elsden, *Biochemistry*, 6 (1967) 2425.
87 B. Smith and C. Franzblau, unpublished results.
88 E. J. Miller, S. R. Pinnell, G. R. Martin and E. Schiffmann, *Biochem. Biophys. Res. Commun.*, 26 (1967) 132.
89 S. Moore, *J. Biol. Chem.*, 238 (1963) 235.
90 R. W. Lent and C. Franzblau, *Biochem. Biophys. Res. Commun.*, 26 (1967) 43.
91 S. M. Partridge, *Federation Proc.*, 25 (1966) 1023.
92 R. U. Lemieux and E. von Rudloff, *Can. J. Chem.*, 33 (1955) 1710.
93 L. L. Salcedo, B. Faris and C. Franzblau, *Biochim. Biophys. Acta*, 188 (1969) 324.
94 M. Tanzer, *Intern. Rev. Connect. Tissue Res.*, 3 (1965) 91.
95 E. J. Miller, G. R. Martin, C. E. Mecca and K. A. Piez, *J. Biol. Chem.*, 240 (1965) 3623.
96 B. L. O'Dell, D. F. Elsden, J. Thomas, S. M. Partridge, R. H. Smith and R. Palmer, *Nature*, 209 (1966) 401.
97 C. I. Levene, *J. Exptl. Med.*, 114 (1961) 295.
98 R. C. Page and E. P. Benditt, *Proc. Soc. Exptl. Biol. Med.*, 124 (1967) 454.
99 B. L. O'Dell, B. C. Hardwick, G. Reynolds and J. E. Savage, *Proc. Soc. Exptl. Biol. Med.*, 108 (1961) 402.
100 N. Weissman, G. S. Shields and W. H. Carnes, *J. Biol. Chem.*, 238 (1963) 3115.
101 B. Starcher, C. H. Hill and G. Matrone, *J. Nutr.*, 82 (1964) 318.
102 D. W. Smith, N. Weissman and W. H. Carnes, *Biochem. Biophys. Res. Commun.*, 31 (1968) 309.
103 L. B. Sandberg, N. Weissman and D. W. Smith, *Biochemistry*, 8 (1969) 2940.
104 J. A. Petruska and L. B. Sandberg, *Biochem. Biophys. Res. Commun.*, 33 (1968) 222.
105 S. R. Pinnell, G. R. Martin and E. J. Miller, *Science*, 161 (1968) 475.
106 M. E. Nimni, *J. Biol. Chem.*, 243 (1968) 1457.
107 C. Franzblau, B. Faris, R. W. Lent, L. L. Salcedo, B. Smith, R. Jaffe and G. Crombie, The chemistry and biosynthesis of the crosslinkages in elastin, in *The Chemistry and Molecular Biology of the Intercellular Matrix*, NATO Advanced Study Institute, 1969.
108 A. J. Bailey, in M. Florkin and E. H. Stotz (Eds.), *Comprehensive Biochemistry*, Vol. 26B, Elsevier, Amsterdam, 1968, p. 297.
109 A. J. Bailey and C. M. Peach, *Biochem. Biophys. Res. Commun.*, 33 (1968) 812.
110 M. L. Tanzer and G. Mechanic, *Biochem. Biophys. Res. Commun.*, 39 (1970) 182.
111 A. Kang, B. Faris and C. Franzblau, *Biochem. Biophys. Res. Commun.*, 39 (1970) 175.
112 A. J. Bailey, L. J. Fowler and C. M. Peach, *Biochem. Biophys. Res. Commun.*, 35 (1969) 663.
113 A. H. Kang, B. Faris and C. Franzblau, *Biochem. Biophys. Res. Commun.*, 36 (1969) 345.
114 H. E. Karrer, *J. Ultrastruct. Res.*, 4 (1960) 420.
115 M. K. Keech, *J. Biophys. Biochem. Cytol.*, 7 (1960) 533.
116 D. C. Pease and S. Molinari, *J. Ultrastruct. Res.*, 3 (1960) 447.
117 D. C. Pease and W. J. Paule, *J. Ultrastruct. Res.*, 3 (1960) 469.
118 K. N. Ham, *Australian J. Exptl. Biol. Med. Sci.*, 40 (1962) 341.
119 W. J. Paule, *J. Ultrastruct. Res.*, 8 (1963) 219.
120 K. Seifert, *Z. Zellforsch. Mikroskop. Anat.*, 60 (1963) 293.

121 W. H. CARNES, W. F. COULSON AND A. M. ALBINO, *Ann. N. Y. Acad. Sci.*, 127 (1965) 800.
122 J. F. WOESSNER JR. AND T. H. BREWER, *Biochem. J.*, 89 (1963) 75.
123 A. H. T. ROBB-SMITH, in G. ASBOE-HANSEN (Ed.), *Connective Tissue in Health and Disease*, Munksgaard, Copenhagen, pp. 15–30.
124 H. E. KARRER, *J. Ultrastruct. Res.*, 5 (1961) 1.
125 K. TAKAGI, *Kumamoto Med. J.*, 22 (1969) 1.
126 K. TAKAGI, *Kumamoto Med. J.*, 22 (1969) 15.
127 R. C. COX AND K. LITTLE, *Proc. Roy. Soc. (London)*, Ser. B, 155 (1961–62) 232.
128 M. D. HAUST, R. H. MORE, S. A. BENCOSME AND J. U. BALIS, *Exptl. Mol. Pathol.*, 4 (1965) 508.
129 W. H. FAHRENBACH, L. B. SANDBERG AND E. G. CLEARY, *Anat. Record*, 155 (1966) 563.
130 A. F. W. HUGHES, *J. Anat.*, 77 (1942–43) 266.
131 H. IWAKUMA, *Arch. Histol. Japan*, 13/1 (1957) 161.
132 K. KINUKAWA, *Arch. Histol. Japan*, 13/1 (1957) 149.
133 M. DYRBYE AND J. E. KIRK, *J. Gerontol.*, 12 (1957) 20.
134 C. SMITH, M. M. SEITNER AND H. P. WANG, *Anat. Record*, 109 (1951) 13.
135 W. Z. SCHWARZ, Z. *Zellforsch. Mikroskop. Anat.*, 63 (1964) 636.
136 E. G. CLEARY, L. B. SANDBERG AND D. S. JACKSON, *J. Cell Biol.*, 33 (1967) 469.
137 T. K. GREENLEE JR., R. ROSS AND J. L. HARTMAN, *J. Cell Biol.*, 30 (1966) 59.
138 J. WAISMAN, W. H. CARNES AND N. WEISSMAN, *Am. J. Pathol.*, 54 (1969) 107.
139 K. TAKAGI AND O. KAWASE, *J. Electron microscopy (Tokyo)*, 16 (1967) 330.
140 S. M. PARTRIDGE, *Biochim. Biophys. Acta*, 140 (1967) 132.
141 O. KAWASE, *Bull. Res. Inst. Diathetic Med., Kumamoto Univ.*, 9 (1959) Suppl. March 10th.
142 R. G. DONOVAN, R. A. ANWAR AND C. S. HANES, Studies of Rheumatoid Disease, *Proc. Third Canadian Conf. Res. in the Rheumatic Diseases*, Univ. of Toronto Press, Toronto, 1965, p. 231.
143 M. MAMMI, L. GOTTE AND G. PEZZIN, *Nature*, 220 (1968) 371.
144 D. W. URRY, B. STARCHER AND S. M. PARTRIDGE, *Nature*, 222 (1969) 795.
145 J. R. BENDALL, *Biochem. J.*, 61 (1955) 31.
146 S. Y. YU AND H. T. BLUMENTHAL, *J. Atheroscler. Res.*, 5 (1965) 159.
147 G. R. MARTIN, E. SCHIFFMANN, H. A. BLADEN AND M. NYLEN, *J. Cell Biol.*, 16 (1963) 243.
148 D. M. KRAMSCH, C. FRANZBLAU AND W. HOLLANDER, *Circulation*, 36, Suppl. 2 (1967) 21.
149 S. GERO, J. GERGELY, T. DEVENYI, L. JAKOB, J. SZEKELY AND S. VIRÁGY, *Nature*, 187 (1960) 152.
150 L. ROBERT, B. ROBERT, M. MOLZAR AND E. MOLZAR, in *Le Rôle de la Paroi Artérielle dans d'Athérogénèse*, Editions de Centre National de la Recherche Scientifique, Paris, 1968, p. 395.
151 M. J. BARNES AND S. M. PARTRIDGE, *Biochem. J.*, 109 (1968) 883.
152 D. M. KRAMSCH AND W. HOLLANDER, *Circulation*, 38, Suppl. 6 (1968) 12.
153 C. FRANZBLAU, F. M. SINEX AND B. FARIS, *Nature*, 205 (1965) 802.
154 K. A. PIEZ, E. WEISS AND M. S. LEWIS, *J. Biol. Chem.*, 235 (1960) 1987.

Chapter XII

The Tubes of Polychaete Annelids

RENÉ DEFRETIN

Zoological Laboratory, University of Lille, Annappes (France)

SUMMARY

The various types of tubes are described. The fact that they appear to be unrelated to the taxonomic position of any particular species is discussed.

(*1*) The organs responsible for tube construction and their function are described, with examples chosen from the Spionidae, Chaetopteridae, Sabellariidae, Sabellidae, Amphictenidae, and Serpulidae.

(*2*) For each representative of the various families, the tube-forming glands and the results of histochemical investigations on the type of secretion are described. These consist chiefly of proteins, mucopolysaccharides, enzymes (peroxidase, EC 1.11.1.7; phenol oxidase), and phenols.

(*3*) In the biochemical section, a description of the methods used is followed by the results of various analyses. These deal with the tubes of *Sabellaria, Spirographis, Sabella, Myxicola, Pectinaria, Lanice,* and *Onuphis.*

Examination of the results obtained shows that the differences between tube types can be explained by marked differences in their composition. Useful information can be gained by histochemical studies on glandular apparatus.

The value of systematic studies is indicated by the interesting but hitherto fragmentary results obtained. Important progress can be achieved by further investigations into the morphology of tube-forming organs, gland histochemistry, and biochemical analysis of the tubes themselves.

1. Introduction

Although many species of polychaete annelids live in secreted tubes, the

[713]

standard classification of polychaetes into sedentary and errant or free-swimming types bears no relation to whether or not a particular species forms a tube.

In the sedentary polychaetes, the body is divided into several regions; the highly modified head is poorly defined, and the gills are localized in a specific zone. They are usually considered as tubicolous. The errant polychaetes are similarly segmented, but the segments just around the mouth form a well-defined cephalic portion, and the terminal segment or pygidium is equally distinct.

While there are many tubicolous species among the sedentary polychaetes, there is no strict correlation between their general morphology and their mode of life. Systematic examination shows that in some cases all the species belonging to whole families secrete no tube, while among other families there may be both tubicolous species and others that are not. References in the literature show that the tubes may be of various types and sometimes of very different composition for representatives of the same family. The tubes may be mucous, membranous, felt-like, encrusted with sand or shell debris, horny, calcareous, opaque or translucent, etc. In true tubicolous species, there are frequently well-differentiated tube-forming organs and an associated glandular system. Lastly, certain species of sedentary polychaetes form no tube, but hollow out burrows whose walls are consolidated or simply lined with a mucus secreted by the worm.

In several families of errant polychaetes there are species which construct a tube, while related species are free-living. The tubes constructed by these errant polychaetes vary in appearance, like those of the sedentary poly-chaetes (they may be mucous, felt-like, membranous, or horny).

Much information on the form and composition of secreted tubes is given by Fauvel[1,2] in his *Faune de France*.

Table I contains a number of examples illustrating both the variability of the appearance of the tubular habitat of certain polychaete annelids, and the apparent independence of tube type on the taxonomic position of the species concerned.

In fact, the protective tubes of polychaete annelids, both sedentary and errant, show no uniformity. Not only are there large variations in mor-phological appearance, composition, and the type of materials incorporated from the environment, but the tubes can also differ in their mode of con-struction and the chemical nature of their constituents. Because of this diversity, consideration of the results obtained on tube composition is

TABLE I

THE TUBES OF POLYCHAETE ANNELIDS

Families	Genera and species	Nature of tube
Ariciidae		No tube.
Spionidae	*Spiophanes bombyx* (Claparède)	Many species secrete no tube.
	Pygospio elegans (Claparède)	Rigid tube of sand.
	Polydora (all species)	Tube of fine sand cemented with mucus. Thin membranous tube covered with clay or mud; certain species hollow out a burrow in chalk or shells.
Magelonidae	*Magelona papillicornis*	No tube, burrows in sand.
Chaetopteridae	*Chaetopterus variopedatus* (Renier)	Membranous tube, supported by parchment-like layers.
	Phyllochaetopterus (all species)	Horny, transparent tube.
Cirratulidae		No tube.
Chloraemidae	*Flabelligera affinis* (Sars)	No tube but sheath of mucus.
Opheliidae		No tube.
Capitellidae		No tube, sometimes burrows lined with mucus.
Arenicolidae		In burrows.
Maldaniidae	*Clymene* (all species)	Arenaceous tube, mostly fragile.
	Johnstonia clymenoides (Quatrefages)	Arenaceous tube.
	Leiochone clypeata (Saint-Joseph)	Tube of fine sand, very fragile.
	Petaloproctus terricola (Quatrefages)	Tube of sand and grit, thick, resistant.
	Owenia fusiformis (Delle Chiage)	Resistant membranous tube, covered with sand.

TABLE I (continued)

Families	Genera and species	Nature of tube
Sabellariidae	Sabellaria alveolata (L.)	Firm tube, made of agglomerated sand grains, forming reefs of Hermella.
Amphictenidae	Pectinaria (Lagis) koreni (Malmgren)	Rigid, fragile tube, made of cemented sand grains.
Ampharetidae		Membranous tube or tube made of agglutinated mud, depending on the species.
Terebellidae	Amphitrite	Several species in burrows.
	Amphitrite affinis (Malmgren)	Thick-walled tube of mud.
	Amphitrite variabilis (Risso)	Fragile tube of sand.
	Amphitrite cirrata (O. F. Müller)	Tube encrusted with mud and sand.
	Terebella lapidaria (Kahler)	Thin tube lining burrows.
	Lanice conchilega (Pallas)	Membranous tube covered with sand and shell debris.
	Nicolea venusta (Montagu)	Thin transparent tube.
	Pista maculata (Dalyell)	Membranous tube covered with sand.
	Thelepus setosus (Quatrefages)	Membranous tube encrusted with grit.
Sabellidae	Sabella pavonina (Savigny)	Elastic tube encrusted with mud.
	Spirographis spallanzanii (Viviani)	Elastic tube encrusted with mud.
	Potamilla reniformis (O. F. Müller)	Horny transparent tube.
	Branchiomomma vesiculosum (Montagu)	Horny tube, thickly encrusted with grit.
	Fabricia sabella (Ehrenberg)	Cylindrical mucous tube.
	Myxicola infundibulum (Renier)	Glairy, thick transparent tube.
Serpulidae	Numerous species of the genera Serpula, Hydroides, Vermiliopsis, Pomatoceros, Salmacina, Apomatus	Opaque calcareous cylindrical or polygonal tube, sometimes embellished with striations and ridges.
	Placeostegus tridentatus (Fabricius)	Translucent or transparent calcareous tube.
	Spirorbis vitreus (Fabricius)	Vitreous translucent tube.
	Spirorbis spirillum (Linné)	Smooth translucent tube.

TABLE I (continued)

Families	Genera and species	Nature of tube
Errant	(The families where all species are free-living are not mentioned.)	
Aphroditidae	Polynoe caeciliae (Fauvel)	Free living except for a few species.
	Panthalis oerstedi (Kinberg)	Mucous tube.
		Felt-like tube.
	Eupanthalis kinbergi (McIntosh)	Tube made up of several felted layers.
Phyllodocidae		Free-living, secreting abundant mucus but not forming tubes.
Syllidae	Eusyllis blomstrandi (Malmgren)	Free-living except for a few species.
	Autolytus edwarsi (Saint-Joseph)	Mucous tubes on laminaria.
		Mucous tube.
	Procerastea halleziana (Malaquin)	Mucous tube in tunic of Ascidia (Ciona).
Nereidae	Nereis pelagica (Linné)	Mucous tube in spurs of laminaria.
	Nereis irrorata (Malmgren)	Membranous tube encrusted with sand.
	Nereis diversicolor (O. F. Müller)	Burrows in sand, consolidated with mucus.
	Platynereis dumerilii (Audouin and M. Edwards)	Mucous or membranous tube.
Eunicidae	Eunice rousseani (Quatrefages)	Free-living except for a few species.
	Eunice siciliensis (Grube)	In burrows hollowed out in sand.
	Onuphis conchylega (Sars)	Burrows in chalk.
	Onuphis quadricuspis (Sars)	Membranous tube covered with shell debris.
	Diopatra neapolitana (Delle Chiage)	Membranous tube encrusted with mud.
	Hyalinoecia tubicola (O. F. Müller)	Membranous tube, thick, leathery.
		Horny transparent tube, resembling a feather quill.

References p. 746

preceded by two sections: one on the specialized tube-forming organs and their function, and one dealing with the cytological and histological findings on the glandular formations responsible.

2. Tube-forming organs

The mucous or membranous tubes constructed by errant polychaetes are produced by the activity of tegmental glands, especially those situated in the parapodial region. *Nereis* offers a particularly good example of this situation. The sand or shell debris sometimes associated with these tubes adheres by direct contact.

The same is true of a large number of sedentary burrowing polychaetes, which live in burrows in rock crevices or burrows hollowed out in sand, or mud. The wall is then consolidated and possibly lined with mucus from the tegmental glands; this is found with certain Spionidae and Maldaniidae. In these instances there is no morphological formation having a mechanical function in tube construction.

However, in those sedentary polychaetes that construct a tube to some extent independent of the substratum, the anterior region possesses lobular formations and glandular complexes whose combined activity participates in tube formation and elongation. Such apparatus can be observed in the Spionidae, Chaetopteridae, Sabellariidae, Amphictenidae, in certain Terebellidae and Sabellidae, and in the Serpulidae.

(a) Spionidae

In *Polydora ciliata*, Dorsett[3] has shown that the lateral lips are used to select sand grains (diameter between 0.03 and 0.05 mm) which are then deposited onto the free border of the tube. The grains are cemented by glandular secretions from the first and second segment. An internal layer of mucus from the glandular pouches of the seventh segment is subsequently applied.

(b) Chaetopteridae

Tube construction has been studied by Barnes[4,5] in various species. In those constructing an annulated tube of horny appearance, a thick glandular epidermis in the anterior region of the body secretes the sheath in semicylindrical sections; the height of a section corresponds to the distance between two annuli. The worm turns itself through an angle of 180° to form the other half-section of the cylinder. In *Ranzanides saggitaria*, the external layer of the tube is made up of agglomerated sand grains; these are collected by the palps and the ventral lip. The particles accumulated on the surface of the ventral lip are added, with the mucus secreted, to the free extremity of the tube. The ventral surface of the worm is subsequently applied to the internal surface of the new formation, and deposits an internal layer of mucus which consolidates the external layer of sand grains.

In the Chaetopteridae, the tube—whether or not embellished with sand grains—is secreted by glands in the ventral surface of the anterior region.

(c) Sabellaridae

The data provided by MacIntosh[6] and Fauvel[2] are ambiguous owing to the existence of some confusion in anatomical terminology. Using reconstruction experiments, Vovelle[7,8] made the following observations on tube construction: The filiform tentacles on the ventral surface of the two opercular peduncles collect shell fragments. At the buccal triangle, those fragments that are too small are eliminated while those of suitable size are transferred downwards by the ventral lips towards the mamilliform protuberances. The latter, which play an essential part in tube construction, make use of two extensions to place each fragment separately onto the border of the pre-existing tube. A protein cement, produced in the form of spherules by the unicellular gland masses, is secreted at the level of the indentations on the inner surface of the mamilliform protuberances; this cement causes shell fragments to adhere. Inside the shell tube an internal sheath, organic and formed in layers, is also produced by a spherular secretion.

(d) Sabellidae

Accurate descriptions of tube construction in the case of *Sabella pavonina* have been published by Nicol[9]. A pair of ventral sacs is observed in the prolongation of the lateral lips surrounding the buccal opening. The mud particles, collected at branchial filament level, pass the lateral lips into the ventral sacs; there they become covered by the secretion of glands opening at the inner surface. Meanwhile, the collar folds deposit secretions from numerous glands onto the upper section of the tube. A string of mucous secretions, sticking the mud particles together, issues from the furrow between the posterior parts of the ventral sacs; it is moulded by the folds of the underlying collar, which apply it to the upper section of the tube while the worm rotates slowly on its axis (Fig. 1).

We ourselves have observed the same process experimentally after introducing Sabellae into glass tubes of appropriate length and diameter. The glass tube was extended by a portion of naturally formed tube; a length of 6 mm was constructed in 3 days. Just as in *Sabellaria*, *Sabella* produces an

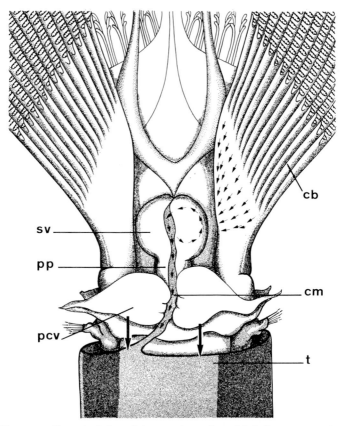

Fig. 1. Diagrammatic ventral view of the anterior region of *Sabella pavonina*; the diagram shows the pathway of particles and the string of mucus during tube elongation (after Nicol); cb, branchial corona; cm, string of mucus; pcv, collar whose ventral folds apply the string of mucus onto the upper section of the tubes, as indicated by arrows; pp, parallel folds; sv, ventral sac; t, tube already formed.

internal organic tube which is almost transparent and contains no extraneous material. Bobin[10] has obtained tube construction experimentally in Sabellae, using the procedure described previously. However, this author finds that the annelid first secretes a mucous tube covering the whole body, which mechanically agglutinates the sand grains.

(e) Amphictenidae

Pallas[11] and MacIntosh[12] have reported some interesting data on *Pectinaria* tubes. However, the most accurate observations on tube growth and mode of formation have been made by Fauvel[13] and Watson[14].

Food particles and sand grains collected by the buccal tentacles are conducted towards the buccal orifice by the horseshoe organ or buccal organ. Sand grains are deposited by the mouth onto the free border of the tube, and are affixed by secretions of the cementing organ situated below the buccal orifice.

In *Pectinaria*, the tube consists of a single layer of sand grains, duplicated internally by a membranous tube formed by several superimposed layers.

(f) Serpulidae

Faouzi[15], Hanson[16], Hedley[17], and Vovelle[18] have provided information on tube construction in *Pomatoceros triqueter*. The collar moulds a muco-calcareous product onto the free border of the tube; the mucous secretion arises from the collar zone, while the calcareous fraction, linked with an organic substrate, is produced by two symmetrical ventrolateral glands below the collar.

In *Eupomatus dianthus*, Muzii[19] describes the internal lumen of the calcareous tube as being lined with an organic membrane; the latter is made up of protein fibers, and probably sulphomucins in the dispersed state.

Ringed structures formed by the Serpulidae during growth have been described by Behrens[20].

Tube construction has also been studied by Soulier[21] in *Branchiostoma vesiculosum*, *Spirographis spallanzanii*, *Sabella viola*, *Protula meilhaci*, *Serpula infundibulum*, and *Hydroides pectinata*; this author discusses the role played by epidermal structures.

These few examples are selected from many comparable instances. They point to the existence of glandular complexes in the anterior region, situated near to and within formations that play a mechanical role in tube construction.

(g) Nereidae

In *Nereis pelagica* and *Nereis irrorata*, which are errant polychaete annelids, there are no specific tube-constructing organs as in the sedentary types. When the worm is withdrawn from its tube, it is found that active secretion of the dorsal parapodial glandular masses rapidly begins to form an elastic tubular coating. During formation, and depending on the nature of the environment, particles of mud, sand grains, or shell debris adhere to it. In *Platynereis massiliensis*, Casanova and Coulon-Roso[22] likewise show that tube construction is brought about by certain parapodia of the anterior region of the body, and that it takes place during the course of complex rotatory and retractile movements of the worm.

3. Glands responsible for tube secretion

The previously mentioned work of Soulier[21] gives numerous descriptions and diagrams of glandular formations in certain species of Sabellidae and Serpulidae. Soulier's investigations were carried out some 80 years ago, and since that time, many publications have appeared on the glandular apparatus of annelids. We shall only consider those cytological and histochemical findings that can be correlated with the biochemical results.

(a) Spionidae

Dorsett[3] has described tube construction in *Polydora ciliata*. The sand grains are cemented by the secretion of glands in the ventral wall of the first and second segment and in the labial epithelium. This secretion is a mucus that hardens rapidly in water. The wrinkled internal wall of the tube thus formed is subsequently smoothed out by deposition of an internal layer of mucus, arising from glandular pouches situated in the seventh segment. This secretion also hardens rapidly. Based on the results of staining with Alcian Blue, the author suggests that these secretions consist of an acid mucopolysaccharide comparable to those identified in the glands and tubes of other polychaetes by Ewer and Hanson[23].

(b) Chaetopteridae

In *Spiochaetopterus oculatus*, Barnes[4,5] observed that the tube is secreted in semicylindrical secretions produced by the activity of epidermal glands. The latter open onto the ventral surface of the cephalic region, at the level of the ninth parapodium. According to the author, the substance of the tube is horny and resembles chitin. No histochemical data are available on the epidermal glands.

(c) Sabellaridae

Vovelle[7,8,24-27] has carried out a detailed study on the glandular secretions of *Sabellaria alveolata*.

(*i*) In the region of the tube-constructing organ there are glands which produce the cementing secretion. This consists of homogeneous and hetero-geneous spherules, whose histochemical reactions are in some ways similar.

Mucopolysaccharide reactions are negative for both categories of spherules. Aldehyde groups can only be demonstrated in the heterogeneous spherules. The protein nature of the spherular secretions is indicated by a series of reactions, but the presence of aromatic radicals is only confirmed for the homogeneous spherules.

There are no sulphydryl groups in the cementing substance. Lastly, the presence of phenolic radicals can be demonstrated in the substance of which spherules are composed. Generally speaking, the homogeneous spherules react more intensely than the heterogeneous ones. The latter might be composed of a stable substance devoid of reactive groups, possibly a protein of already stabilized structure.

(*ii*) On the ventral surface, the first two parathoracic segments possess rod-containing glandular cells. The secretion is a sulphydryl-containing mucoprotein; all the reactions for acid and neutral mucopolysaccharides and for sulphydryl containing proteins are positive. This secretion also contains a polyphenol oxidase enzyme.

(*iii*) The ventral surface of the third parathoracic segment possesses glandular cells. The alcohol-soluble secretion of these cells gives all the reactions of an *o*-diphenol.

(*iv*) Between these last two glandular categories, there are mucocytes whose secretion consists of acid mucopolysaccharides; these form a buffer zone between the phenols and phenolase.

References p. 746

(*v*) The posterior abdominal cirri possesses cells with a protein secretion associated with a DOPA-oxidase; these cells might participate in posterior occlusion of the tube.

(*vi*) Mucocytes are present on all the parapodia, and produce ordinary secretion of epidermal mucus. According to the author, when the labile protein containing phenolic radicals (homogeneous spherules) is secreted, it is exposed to the action of a polyphenol oxidase (EC 1.10.3.1) (rod-containing glandular cells) and an *o*-diphenol (cells with an alcohol-soluble secretion). Dissociation of the heterogeneous spherules could give rise to scleroprotein platelets which are agglutinated with the labile protein tanned by the concomitant action of *o*-diphenol and polyphenol oxidase.

(*d*) Amphictenidae

According to Fauvel[13], in *Pectinaria* tubes an external layer formed of sand grains stuck together by a cement and several internal membranous layers can be distinguished. The cement shows no affinity for mucin-staining reagents, while the internal layers react positively.

Defretin[28] has observed hyaluronic acid-rich cells in the cementing organ described by Watson[14]. The ventral glandular protuberances, which according to Watson[14] secrete the substance forming the internal layers of the tube, are poor in hyaluronic acid and the clear ovoid glands situated just behind the first pair of nephridia are totally devoid of hyaluronic acid. They secrete an abundant glairy mucin, which assists manipulation of sand grains by the great golden chaetae projecting from the head. This secretion might possibly participate in extending the tube's small extremity.

(*e*) Terebellidae

In *Lanice conchilega*, Defretin[29] has observed three glandular categories:

(*1*) (*a*) *cells*: the secretion consists of small spherules, is not stained by mucicarmin, is not metachromatic, and gives no positive reaction in tests for polysaccharides; the spherules appear to be protein in nature;

(*2*) (*b*) *cells*: these produce a fine lamellar secretion, which is stained with mucicarmin, is metachromatic, and has a high content of polysaccharides and hyaluronic acid;

(*3*) (*c*) *cells*, whose frothy secretion has the same histochemical characteristics as the (*b*) cells, but to a smaller extent, while hyaluronic acid can be detected in some cases (Fig. 2a, 2c).

The region responsible for the tube construction (upper lip, lateral lobes) contains the (a) and (c) cells; and hyaluronic acid is detectable in the latter type. The epidermal glands of the anterior thoracic region possess the same cell types—(a) and (c)—but do not appear to contain hyaluronic acid.

The large ventral gland mass (clypeal tissue) shows all three types of gland cell [(a), (b), and (c)]; Courtois[30] has demonstrated peroxidase activity in this tissue, using Lison's zinc–leuco dye method and Graham's procedure.

To sum up, protein substances, mucopolysaccharides, hyaluronic acid, and peroxidases have been identified in the glandular secretions of zones participating in tube construction.

(f) Sabellidae

In *Spirographis spallanzanii* and *Sabella pavonina*, numerous glands open onto the internal surface of the ventral sacs; their secretion participates in forming a string of agglutinated particles which is coiled onto the upper section of the tube. The lower surface of the collar is particularly rich in glandular formations; these are involved in the collar's role in tube construction.

In these species it is also possible to observe the presence of two main types of glandular cells; one type is azurophilic, and the others fuchsinophilic. The azurophil cells secrete a true mucin which gives a positive reaction to the following techniques: PAS, Alcian Blue, mucicarmine, and the Hale and Kulonen methods for detecting hyaluronic acid. The fuchsinophil cells have a protein secretion.

In *Myxicola infundibulum*, there is no tube-constructing apparatus; the mucous tube, which is thick, solid and translucent, is produced by the activity of numerous epidermal glandular cells. Their secretion gives positive mucopolysaccharide reactions; it is rich in hyaluronic acid.

(g) Serpulidae

In *Pomatoceros triqueter*, Hedley[17] showed that the glandular secretions playing a part in tube construction come from two formations: one pair of calciferous glands arranged on either side of the peristomium, and a large mass of mucous cells in the ventrolateral epithelium of the peristomium. The presence of calcium in the former is masked by an organic substance which inhibits the reactions for detecting calcium. Its presence can, however,

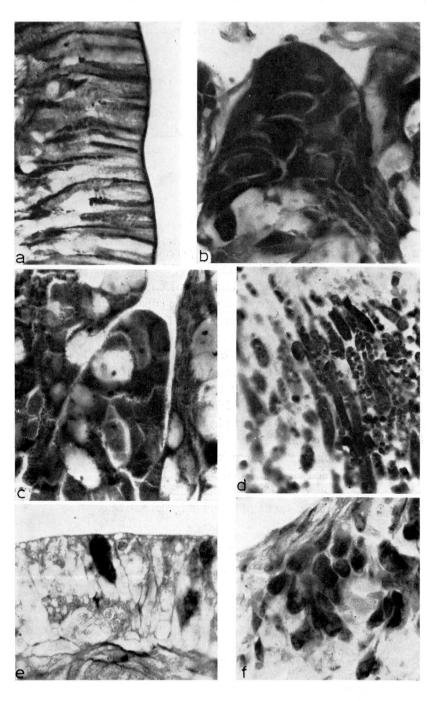

be demonstrated after incineration. Histochemical reactions show that the mucous cell secretion contains a sulphomucopolysaccharide and an appreciable proportion of calcium. The secretions of both glandular categories, present in the fold of the peristomial collar, are moulded by the collar over the anterior border of the tube. It is possible that when the two secretions are mixed, the mucoitin sulphuric acid reacts with calcium cations to form a complex.

The calcareous material remains pliable during deposition, and subsequently hardens. This hardening of the secretion is found not only in the Serpulidae, but has also been observed by various authors in tubes formed exclusively by the secretion of glands producing sulphomucopolysaccharides. Hyaluronic acid might be responsible for this hardening: it has been detected in the glands of other Serpulidae, such as *Protula intestinum* and *Vermiliopsis infundibulum* (Fig. 2e).

(h) Nereidae

The substance used for tube construction seems to come primarily from the glandular masses situated along the dorsal border of the parapodia. These glandular formations are considerably developed in tubicolous nereids, particularly in *Nereis pelagica* and *Nereis irrorata*. The cytological and histochemical investigations of Defretin[31-35] have provided the following results.

Three glandular categories are distinguished: (*i*) the tegmental glands in the dorsal and ventral epidermis, (*ii*) the fuchsinophil glands constituting the dorsal parapodial masses, (*iii*) the fine granule glands of the parapodial projections and the thoracic region.

Metachromatic azurophil tegmental glands stain with mucicarmine and are PAS-positive; they secrete mucopolysaccharides. A phenol oxidase is

---

Fig. 2. (a) Excretory canals of the anterior thoracic epidermal glands of type (*c*) in *Lanice conchilega*; note the high content of polysaccharides demonstrated by Hotchkiss technique (PAS), × 475. (b) Ventral glandular mass (clypeal tissue) of *Lanice conchilega*, rich in polysaccharides, Hotchkiss technique (PAS), × 475. (c) Cells of type (*b*) in clypeal glandular tissue of *Lanice conchilega*, detection of hyaluronic acid, Hale's technique, × 475. (d) Spherule-containing fuchsinophil glands of the dorsal parapodial masses in *Nereis irrorata*, detection of hyaluronic acid, Hale's technique, × 475. (e) Epidermal glands of *Protula intestinum*, detection of hyaluronic acid, Kulonen's technique (hyaluronidase followed by AgNO₃), × 475. (f) Glands of anterior segments in *Hyalinoecia tubicola*, presence of hyaluronic acid, Hale's technique, × 475.

also associated with the secreted mucin, which is devoid of hyaluronic acid. The fuchsinophil glands, which are long and tubular, are saturated with secretion spherules; they react negatively to all the polysaccharide and mucopolysaccharide tests (mucicarmine, PAS, metachromasia, etc.). Hyaluronic acid can, however, be detected (Fig. 2d). They are stained by ninhydrin, and the small spherules in the proximal zone of the gland are rich in peroxidase. The fine granule glands are also resistant to mucopolysaccharide reagents; these glands, which make up large thoracic masses in certain tubicolous nereids, have a protein-type secretion like the previous ones.

These observations on the mode of tube construction and histochemical characteristics as summarized above suggest that the highly elastic tube of these nereids is composed chiefly of protein. Since a peroxidase and a polyphenol oxidase can be detected in the glandular apparatus, it is possible that the tube is similar in composition to that of the *Sabellaridae*. However, before confirming the existence of a tanned protein, the presence of phenol in one category of the glands concerned must first be demonstrated.

(i) Eunicidae

In *Hyalinoecia tubicola*, which secretes a tube of horny appearance, the anterior segments show large glands descending into the coelom; their orifices open at the level of the lateral ventral protuberances. The presence of hyaluronic acid has been detected by Defretin[28] in certain glands (Fig. 2f).

SUMMARY

Histochemical examination of the glands taking part in tube construction reveal the existence of many cellular categories. In most cases the tube appears to be composed of a material formed from several secretions produced by different glandular cells.

This has been conclusively demonstrated by Vovelle, as mentioned earlier, for *Sabellaria alveolata*, and by Hedley for *Pomatoceros triqueter*. The same is probably true for many species.

In view of the results obtained hitherto, several hypotheses can be considered: (*1*) The presence of protein secretions, phenolic secretions, and phenol oxidase suggests that a tanned protein may be formed. (*2*) The concomitant existence of protein secretions, sulphomucopolysaccharides and sometimes of hyaluronic acid are fairly indicative of mucoprotein formation. (*3*) The low content or absence of protein secretion, the marked development of mucopolysaccharide glands, and the presence of hyaluronic acid, as found in *Myxicola infundibulum*, provide evidence for the formation of high-viscosity mucin.

The problems arising can only be solved by detailed and exhaustive histochemical investigations on each type of glandular cell in the various species of tubicolous polychaetes.

4. Biochemical investigation of the tubes

The dry tubes are powdered, and the content of dry material and the percentage of water can then be determined by drying in an oven. The inorganic salts and low-molecular organic substances (excretion products) are separated from the protein part of the tubes by dialysis. Centrifugation of the non-dialysable material yields: (*a*) an insoluble fraction, which is washed, dried, and weighed, and (*b*) a soluble fraction, which is concentrated and lyophilized.

To study the glycoproteins in these two fractions, the protein part and the polysaccharide part are analysed separately.

I. METHODS

(*a*) *Protein part*

(*i*) *Separation of the amino acids*

The protein may be isolated by the action of soda (Neuberg–Cahill method). The protein is then precipitated with acetic acid, while the supernatant liquid contains the carbohydrate prosthetic group. After dialysis, the protein is hydrolysed with hydrochloric acid in sealed tubes (5.6 N HCl at 100°, 36 h). This leads to total breakdown into amino acids, and the hydrogen chloride is removed under vacuum (Defretin[36]).

However, it is not essential to separate the protein from the carbohydrate prosthetic group, and the hydrolysis with hydrochloric acid may be carried out directly on the powdered tubes. This was the method used in our work with Drugy and Montreuil. The hydrogen chloride is then eliminated by stirring the hydrolysis mixture with Permutit 50. The amino acids retained on the resin are displaced by ammonia solution after the resin has been washed. The ammonia solution is evaporated to dryness.

It should be noted that, despite the dialysis that precedes these operations, the non-dialysable material still contains a large proportion of iron salts, which cause partial destruction of amino acids during the acid hydrolysis. The amino acid determinations are accordingly incorrect.

(*ii*) *Identification of the amino acids*

The dry residue is taken up in distilled water and analysed by two-dimensional chromatography with the following solvent systems: (*1*) *n*-butanol–acetic acid–water (4:1:5 by volume), (*2*) phenol saturated with water in an ammoniacal atmosphere (3% ammonia). The amino acids are detected with ninhydrin.

(*b*) *Carbohydrate part*

The carbohydrates forming the glycoproteins of the tubes can be determined.

(*i*) *Carbohydrates*

The total carbohydrates are determined by Rimington's modification[37] of the Tillmans–Philippi method[38]. The results obtained by this colorimetric method using orcinol must, however, be corrected to allow for the quantity of uronic acids contained in the sample.

(*ii*) *Glycosamines*

The colorimetric determination of the glycosamines with the modified Ehrlich reagent (*p*-dimethylaminobenzaldehyde dissolved in concentrated hydrochloric acid) is carried out by the method described by Belcher *et al.*[39], which is itself a modification of the original method of Elson and Morgan[40].

(*iii*) *Uronic acids*

The uronic acids are determined colorimetrically by Dische's method[41], which is based on the purple colour produced by carbazole in sulphuric acid solution.

(*iv*) *Sialic acids*

The sialic acids can be determined by the method of Werner and Odin[42] using Dische's diphenylamine reagent.

(*v*) *Separation of the carbohydrates*

Defretin[43] isolated the carbohydrate prosthetic group with soda by the method described by Neuberg and Cahill[44], and then hydrolysed it with

0.5 N sulphuric acid in sealed tubes at 100° for 4 h. The sulphuric acid was eliminated with barium hydroxide. In more recent work with Drugy and Montreuil, we hydrolysed the powdered tubes directly with 2 N hydrochloric acid at 100° for 2 h. The hydrolysis product was purified by passage over Dowex-50 and then over Duolite A 40.

(*vi*) *Identification of the monosaccharides by paper chromatography*
Defretin[43] carried out two-dimensional chromatographic analyses with the following solvent systems: (*1*) *n*-butanol–acetic acid–water (4:1:5 by volume); (*2*) phenol saturated with water in an ammoniacal atmosphere (30% ammonia).

One-dimensional chromatography in the first system has also been used. The following reagents have been used for detection: phthalic acid–aniline in water saturated with butanol, benzidine dissolved in acetic acid–ethyl alcohol, and ammoniacal silver nitrate.

Drugy used the following solvent systems for chromatographic analyses: (*1*) *n*-butanol–acetic acid–water (4:1:5 by volume); (*2*) pyridine–ethyl acetate–water (1:2:2 by volume).

The specific reagents used by this author were Partridge's aniline oxalate reagent[45] for the aldoses, and Dedonder's urea hydrochloride reagent[46] or Johanson's anthrone reagent[47] for the ketoses.

II. RESULTS

(*a*) *Sabellaridae*

A detailed study of the construction of the tube of *Sabellaria* and of the glandular organs responsible has been carried out by Vovelle[8]. Table X of this author's article contains not only the results of his histochemical investigations on the secretion of the glandular cells involved, but also an interpretation of the role of each type of gland. However, he does not appear to have used biochemical analyses, which are the only means of obtaining information about the constituents. Vovelle was nevertheless able to show that the homogeneous protein spherules from deep glandular formations emerging at the level of the building organ form the raw material of the organic cement of the tube; on the other hand, two parathoracic segments secrete a mucoprotein having SH groups and a polyphenol oxidase, while the following segment produces an alcohol-soluble *o*-diphenol.

The homogeneous substance of the spherules (protein containing phenolic groups) appears to associate with a quinone resulting from the action of a phenolase on a phenol. This would give a tanned protein, as shown in the following scheme, which is responsible for stability of the tube.

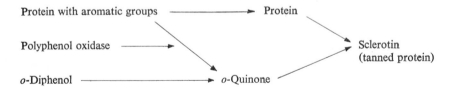

Moreover, the heterogeneous protein spherules, which are twice as numerous as the homogeneous spherules, dissociate during their emission. Their inclusions, which consist of a structural protein, and which are liberated in the form of platelets, agglutinate with the freshly tanned labile protein. They participate in the formation of the matting observed on examination of tube sections in the electron microscope.

This interpretation fits into the general scheme of the production of structural proteins by tanning with quinones in certain invertebrates, as shown by Pryor *et al.*[48] for the ootheca of the cockroach, by Hackmann[49] for the cuticle of insects, and by Smyth[50] for the eggshell of helminths and the byssus of the mussel.

There is no doubt that a biochemical study of *Sabellaria* tubes would allow the elimination of the uncertainties concerning the precise nature of the aromatic groups. Chromatographic analysis, by revealing the amino acid composition of the protein, would provide the information that cannot be derived with certainty from Vovelle's histochemical study.

(b) Amphictenidae

The tube of *Pectinaria koreni*, which is formed from grains of sand bound together by a cement having a bubbly structure, has been studied by Lafon[51]. The composition of the constituents of the internal tube, which consists of several layers, has been reported by Defretin[36,44]. The cement accounts for 11.9% of the dry weight of the tubes. It retains its structure after ashing, pointing to a combination of organic and inorganic constituents. After extraction with acid or alkali, more than half of the ashed cement is found to consist of inorganic substances. Lafon's analyses show that the inorganic components are silicates of aluminium and calcium and masked iron combined with organic components. According to this author, the organic components have a nitrogen content of 12 %, which he attributes to protein, and a sulphur content of 5 %, one-quarter of which is in the oxidized state (inorganic sulphates or organosulphuric compounds) and three-quarters corresponds to the amino acids containing sulphur in the proteins of the cement; they also contain 5 % of reducing carbohydrates liberated by the acid hydrolysis of the glycoproteins. Lafon deduced that the organic part, which forms less than half of the cement, is glycoprotein

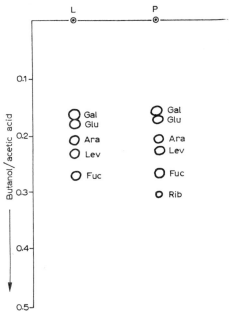

Fig. 3. One-dimensional chromatogram of the carbohydrate fraction of tubes of several polychaetes: L, *Lanice conchilega*; P, *Pectinaria koreni*. The unidentified sugars are not shown on this diagram; they are shown for the Sabellaridae, Fig. 5.

having a low carbohydrate content but a high content of neutral sulphur.
On chromatographic analysis of the carbohydrate fraction of the internal part of the tube of *Pectinaria*, Defretin was able to detect the presence of galactose, glucose, arabinose, levulose, fucose and ribose, as well as 5 unidentified carbohydrates that are also found in the tubes of other species and whose nature is discussed in connection with the constitution of the

tubes of Sabellidae (Fig. 3). However, these compounds appear to be present in much smaller quantities than the sugars that have been identified.

Chromatographic analysis of the protein fraction allows the identification of the following amino acids: aspartic acid, glutamic acid, serine, glycine, threonine, valine, methionine, leucine, and phenylalanine (Fig. 4). The diamino acids and proline, which have frequently been found in other species, have not been detected in the tube of *Pectinaria*. However, the analyses were carried out on small quantities of tubes, and should be repeated with larger samples to allow identification of the amino acids that are present in small proportions.

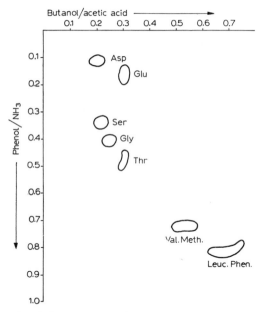

Fig. 4. Two-dimensional chromatogram of the amino acids from the protein fraction of a tube of *Pectinaria koreni*.

These analyses should be applied separately to the outer tube, consisting of cemented grains of sand, and to the membranous inner tube. They would allow the confirmation of Lafon's report that the organic sulphur corresponds mainly to amino acids containing sulphur in the protein part of the cement.

(c) Terebellidae

Lafon[51] showed for *Lanice conchilega*, as for *Pectinaria*, the existence of an organic–inorganic cement forming the outer fibrous sheath, on which are laid grains of sand and shell debris. The inorganic components are silicates, but these are different from the silicates of *Pectinaria*. They have lower contents of calcium (11% instead of 21%) and alumina (6% instead of 18%), whereas they are rich in silica (60% of the ash) and the proportion of iron is much smaller (2.5% instead of 11.5%). Analysis of the organic fraction shows less nitrogen than in *Pectinaria* (9.5% instead of 12%), but a higher sulphur content (7% instead of 5%) and a large proportion of carbohydrates, which account for one-third of the organic material.

Lafon concludes that the mucoprotein of the cement of *Lanice conchilega* is characterized by the abundance of carbohydrate and sulphonic prosthetic groups, the greater part of the sulphur being in the form of compounds such as mucoitin sulphuric acid. The high content of carbohydrates agrees with the qualitative diversity reported by Defretin, who identified galactose, glucose, arabinose, levulose, fucose, probably mannose, and also obtained four unidentified carbohydrates similar to those found for *Pectinaria* (Fig. 3).

The same author reported that the protein fraction contains a large quantity of aspartic acid, glutamic acid, serine, and glycine, a medium proportion of alanine, the valine–methionine group, and the leucine–phenylalanine group, a smaller content of proline, histidine, lysine, and arginine, and very little tyrosine. These estimates are merely based on the size and optical density of the spots on the chromatograms.

Further investigations on the constituent amino acids could be usefully directed towards the identification of the amino acids containing sulphur. In view of the high sulphur content reported by Lafon, it is important to establish whether it is all present as mucoitin sulphuric acid or whether some of it is involved in the protein structure.

(d) Sabellidae

The first results of chromatographic analyses on the construction of the tubes of *Spirographis spallanzanii*, *Sabella pavonina* and *Myxicola infundibulum* were provided by Defretin *et al.*[36,36a,43]. This work has been resumed since then by Drugy, Montreuil and Defretin*, using a wider range of experimental techniques, which have been described above. It should be noted that the investigations were carried out: (*a*) on the complete tube of *Myxicola* (this tube having a homogeneous structure); (*b*) on the entire tube of *Sabella* on the one hand, and on its upper part on the other (this tube having a small lower zone of different appearance), and (*c*) on the entire tube of *Spirographis* and on a tube that had been carefully scraped before drying (this tube being heavily coated with mud particles on the outside).

The results obtained refer to these peculiarities; they relate (*i*) to the distribution of dialysable and non-dialysable components, (*ii*) to the carbohydrate composition, and (*iii*) to the amino acid composition.

(i) Dialysable and non-dialysable components

The results are shown in Table II. The tubes of *Myxicola* have a very high content (81.55%) of dialysable substances, probably inorganic salts. The tubes of these annelids contain large quantities of soluble proteins, sometimes as much as 20% by weight of the entire non-dialysable fraction. The complete tubes of *Spirographis* contain 4 times as much dialysable material as tubes that have previously been scraped.

TABLE II

WEIGHT ANALYSIS (expressed in g per 100 g) OF TUBES OF *Myxicola, Sabella* AND *Spirographis*

Species	Part of tube	Moisture	Not dialysable, insoluble	Not dialysable, soluble	Not dialysable, total	Dialysable (calculated)
Myxicola	Complete	3.4	13	5.45	18.45	81.55
Sabella	Complete	2	98	traces	—	—
	Upper	0	87	5.6	92.6	7.4
Spirographis	Complete	0	71	1.9	72	28
	Complete (scraped)	8	75	19.1	94	6

* F. Drugy, J. Montreuil and R. Defretin, unpublished work carried out in the Biological Chemistry Laboratory of the Faculté des Sciences de Lille (Director: Prof. J. Montreuil) and at the Institut de Biologie Maritime de Wimereux (Director: Prof. R. Defretin).

TABLE III

SUGAR COMPOSITION OF THE NON-DIALYSABLE SOLUBLE AND INSOLUBLE FRACTIONS OF THE TUBES OF *Myxicola, Sabella* AND *Spirographis*

(g per 100 g of dry substance)

Species	Soluble fraction						Insoluble fraction					
	Carbo-hydrates	Uronic acids	Glycos-amines	Carbohy-drates/ uronic acids	Carbohy-drates/ glycos-amines	Uronic acids/ glycos-amines	Carbo-hy-drates	Uronic acids	Glycos-amines	Carbohy-drates/ uronic acids	Carbohy-drates/ glycos-amines	Uronic acids/ glycos-amines
Myxicola total tube	16.47	11.7	3.56	1.40	4.62	3.28	14.72	9.1	3	1.61	4.90	3.03
Sabella total tube upper part	— 6.92	— 3.05	— 2	— 2.26	— 3.46	— 1.53	6.84 7.02	3.05 3.20	2.1 2	2.24 2.20	3.26 3.51	1.45 1.60
Spirographis total tube scraped tube	11.45 10.80	— 4.2	3.12 2.55	— 2.57	3.67 4.20	— 1.64	5.15 8.29	3.37 5.60	1.30 3.85	1.82 1.48	3.90 2.15	2.60 1.47

References p. 746

(ii) Examination of the carbohydrate fraction

The specific colour reactions show the presence of neutral sugars, glycosamines, and uronic acids in the non-dialysable soluble and insoluble fractions. Sialic acids and inositol are absent. Colorimetric determination of the carbohydrate components gives the following results (Table III). Paper chromatography of the total mucoid carbohydrates from the non-dialysable soluble and insoluble fractions of the same annelids (Fig. 5) allowed the determination of their proportions (Table IV).

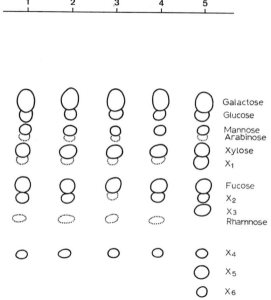

Fig. 5. Chromatogram of sugars in the neutral carbohydrate fraction of tubes of *Sabella*, *Spirographis* and *Myxicola*. 1, Whole tube of *Sabella*; 2, upper part of *Sabella* tube; 3, whole tube of *Spirographis*; 4, scraped tube of *Spirographis*; 5, whole tube of *Myxicola*. Solvent system: *n*-butanol–acetic acid–water (4:1:5 by volume); visualized by oxalic acid–aniline reagent; X_1, X_2, X_3, X_4, X_5, X_6: unidentified sugars.

The interpretation of these results leads to the following observations: (*a*) The tubes of these annelids contain large proportions of carbohydrates (about 30% in *Myxicola* and about 15% in *Sabella* and *Spirographis*). (*b*) The compositions of the tubes of *Sabella* and *Spirographis* are very similar, whereas the tubes of *Myxicola* are richer in carbohydrates. (*c*) The carbohydrate compositions of the soluble fractions and of the insoluble fractions are very similar. (*d*) The uronic acid content is very high in the case of *Myxicola*. (*e*) There appears to be no appreciable difference between the upper part of the tube of *Sabella* and the complete tube (including the lower part of different appearance). (*f*) The galactose content of the poly-

TABLE IV

SUGAR COMPOSITION OF THE TOTAL MUCOIDS OF THE NON-DIALYSABLE SOLUBLE AND INSOLUBLE FRACTIONS OF THE TUBES OF *Myxicola, Sabella* AND *Spirographis*

Neutral sugars	R galactose[a]	Colour of spot[b]	Myxicola	Sabella		Spirographis	
				Complete tube	Upper part	Complete tube	Scraped tube
Galactose	1.00	B	30	30	30	30	30
Glucose	1.12	B	10	10	10	10	10
Mannose	1.27	B	2	2	2	2	2
Arabinose	1.37	R	1	1	1	1	0
Xylose	1.53	R	15	15	10	10	10
X_1	1.63	B	20	1	1	0	0
Fucose	1.80	B	10	10	10	8	10
X_2	1.92	B	2	3	3	0	2
X_3	2.10	R	8	traces	traces	traces	traces
Rhamnose	2.14	B	traces	traces	traces	traces	traces
X_4	2.52	R	1	3	3	0	2
X_5	2.74	B	3	0	0	0	0
X_6	3.00	R	1	0	0	0	0

[a] Migration velocity of the sugar/migration velocity of galactose.
[b] Detection with aniline–oxalate reagent. B, brown; R, red.

saccharide fraction is very high for all three species. (*g*) The sugars X_1 to X_6 reported earlier by Defretin[43] have been detected; however, it has not been possible to study them, since they have not been isolated in sufficiently large quantities. Nevertheless, their chromatographic migration rate suggests that they are methylpentoses and bis-deoxyhexoses.

(*iii*) *Examination of the protein fraction*
    Hydrolysis of the proteins forming the annelid tubes with hydrochloric acid followed by chromatography and electrophoresis shows the presence

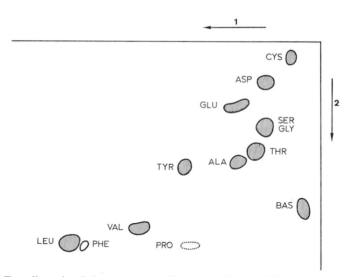

Fig. 6. Two-dimensional chromatogram of the hydrochloric acid hydrolysate of the total protein from *Sabella*. Direction 1: *n*-butanol–acetic acid–water (4:1:5 by volume). Direction 2: phenol saturated with water, ammoniacal atmosphere, detection with ninhydrin reagent.

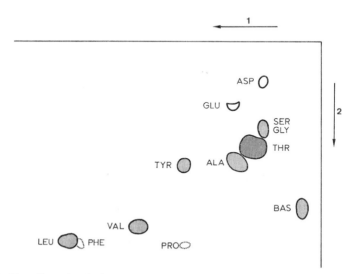

Fig. 7. Two-dimensional chromatogram of the hydrochloric acid hydrolysate of the soluble proteins from *Spirographis*. Direction 1: *n*-butanol–acetic acid–water (4:1:5 by volume). Direction 2: phenol saturated with water, ammoniacal atmosphere, detection with ninhydrin reagent.

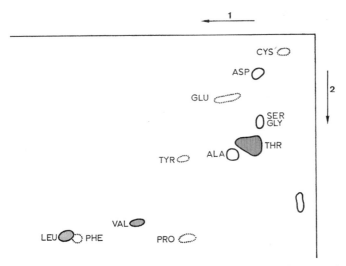

Fig. 8. Two-dimensional chromatogram of the hydrochloric acid hydrolysate of the total proteins from *Spirographis*. Direction 1: *n*-butanol–acetic acid–water (4:1:5 by volume). Direction 2: phenol saturated with water, ammoniacal atmosphere, detection with ninhydrin reagent.

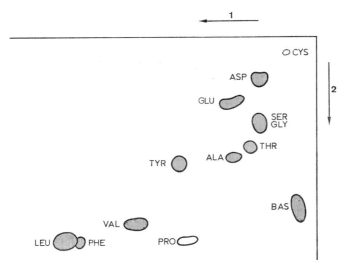

Fig. 9. Two-dimensional chromatogram of the hydrochloric acid hydrolysate of the soluble proteins from *Myxicola*. Direction 1: *n*-butanol–acetic acid–water (4:1:5 by volume). Direction 2: phenol saturated with water, ammoniacal atmosphere, detection with ninhydrin reagent.

References p. 746

TABLE V

AMINO ACID COMPOSITION OF THE TOTAL PROTEINS AND OF THE SOLUBLE PROTEINS OF THE TUBES OF *Myxicola, Sabella* AND *Spirographis* (per 10 alanine residues)

Amino acids	Myxicola (soluble proteins)	Sabella (total proteins)	Spirographis Total proteins	Spirographis Soluble proteins
Aspartic acid	4	15	15	4
Glutamic acid	2	15	15	4
Arginine⎱ Histidine⎬ Lysine ⎰	20	15	20	20
Alanine	10	10	10	10
Cysteine	1	10	10	1
Glycine	18	18	18	18
Leucine	15	20	20	25
Phenylalanine	4	6	6	4
Proline	+	++	+++	+
Serine	12	12	12	20
Threonine	45	25	25	50
Tyrosine	8	6	10	12
Valine	15	15	15	25

of numerous amino acids (Figs. 6–9, Table V). The amino acids and their relative proportions suggest the following: (*a*) The total proteins of tubes of *Sabella* and of *Spirographis* have very similar compositions, which differ appreciably from the total protein composition of the *Myxicola* tube. (*b*) The threonine content is very high for the three annelids in question, particularly in the soluble protein fraction. (*c*) Cysteine is present in a considerable proportion in the total proteins, but in a very small proportion in the soluble proteins. (*d*) For these three *Sabellidae* the proportion of dicarboxylic amino acids is small, whereas there is a large quantity of basic amino acids, among which lysine predominates.

(*e*) *Discussion on the tubes of the Sabellidae*

These results add further information to that provided by the earlier investigations of Defretin *et al*. They emphasize the sometimes large quantity of inorganic substances with which the tube is impregnated, particularly in *Myxicola*. They show that the tubes examined have a heterogeneous constitution. Part of the protein is denatured while another part has remained "native", probably because of its high content of carbohydrate. It is the second, water-soluble fraction that can be separated from the denatured fraction by aqueous extraction. This soluble fraction is also very poor in cysteine, whereas the total proteins are rich in this sulphur-containing amino acid. This heterogeneous composition undoubtedly corresponds to two parts in the tube structure. One, consisting of sulphur-rich scleroproteins, forms the solid skeleton of the tube. The other, which consists of glycoproteins, is trapped in the meshes of the scleroprotein structure as the tube is being formed.

Investigation of the glycoprotein fraction shows that it is not a mucoprotein formed by the association of a protein with an acidic mucopolysaccharide, like hyaluronic acid. If this were so, the uronic acid content should be smaller than or equal to the glycosamine content, whereas the polysaccharide fraction of the glycoproteins of the tubes is richer in uronic acids than in glycosamines. However, the nature of the bonding of the uronic acid still remains to be established.

The difference between the compositions of the tubes of *Myxicola* on the one hand and of *Sabella* and *Spirographis* on the other should be emphasized: 32% of total carbohydrates in the first case, and 12–17% in the second. Moreover, the glycoproteins of the *Myxicola* tubes are richer in unidentified sugars, particularly X_1, X_3, X_5 and X_6. The protein fraction of the *Myxicola* tube contains less glutamic acid, leucine, serine, tyrosine, and valine than that of the other two species.

(f) Eunicidae

On partition paper chromatography of the total hydrolysate of the tubes of *Hyalinoecia (Onuphis) tubicola*, Defretin[36] found an amino acid composition similar to that obtained for *Spirographis spallanzanii*. It contained high serine and glycine contents, medium contents of aspartic acid, glutamic acid, alanine, valine and leucine–phenylalanine, small quantities of lysine, histidine and arginine, and traces of threonine and proline. These estimates are based on the optical density of the spots on the chromatogram.

The hydrolysate of the carbohydrate prosthetic group shows the presence of glucosamine.

As early as 1882, Schmiedeberg[52] had suggested that the tube of this annelid consisted of a carbohydrate "onuphin", an albuminoid substance, and acid calcium and magnesium phosphates.

Graham et al.[53] showed more recently that the whole of the phosphate extracted by 1 N HCl is non-dialysable, the phosphate macromolecule comprising 70% by weight of the tube. This polymer contains 13.5% of phosphorus and less than 0.5% of nitrogen. On hydrolysis followed by chromatography, these authors found aldohexose and aldohexose phosphate, but did not detect either uronic acid or glucosamine. The non-dialysable substance obtained by extraction with acid appeared to be a stable compound of phosphate and hexose having a high molecular weight and containing a low proportion of nitrogenous substances. The name "onuphic acid" was proposed.

These data do not seem to conflict with the results reported by Defretin et al. Graham et al. identified onuphic acid in the acid extract and observed that the tube forms a gelatinous mass on demineralization. They found that, after washing, this mass contains only a small proportion of phosphate, but they do not appear to have investigated its composition. Defretin et al. hydrolysed the tube in 6 N HCl in a sealed tube at 100°. Their results, which were outlined above, refer to the material of the tube as a whole, i.e. the gelatinous mass plus the acid extract of Graham's work. The gelatinous mass probably corresponds to the albuminoid substance reported by Schmiedeberg.

Without a complete analytical study, it can be concluded from a comparison of the results reported by the above authors that the tube of Hyalinoecia tubicola consists of a combination of onuphic acid and a mucoprotein.

5. Conclusions

The results of the investigations on the tube formation, the secretory glands, and the chemical composition of the tubes, though fragmentary, draw attention to the diversity of the building apparatus, to the histochemical nature of the glandular secretions, and to the marked differences in the chemical compositions of the tubes.

The glandular secretions vary, and consolidation of the substances secreted appears to take place by different processes. In some cases it results from the tanning of a protein (Sabellaria), while in others a glycoprotein secretion is associated with hyaluronic acid (Lanice, Nereis and others). Thus a very thorough histochemical study of the glandular cells concerned will be necessary in order to establish the mode of consolidation of the tube material in each case.

Although the results of the analyses carried out so far are few in number, notable differences have been found. Special substances are present in certain cases, such as onuphic acid in *Hyalinoecia*. Attention should, however, be drawn to the similarity of composition found for similar species secreting similar tubes; this is so *e.g.* for *Spirographis* and *Sabella*. In another related family, on the other hand, *Myxicola* is characterized by a tube that is quite different in its composition, richer in carbohydrate and poorer in protein.

Our knowledge of the tubes secreted by the polychaete worms is still very limited, but in many cases their structure contains an association of a protein and a carbohydrate. It appears that a combination of a detailed histochemical study of the glands and the complete biochemical analysis of the tubes is necessary in each individual case in order to obtain significant results.

References p. 746

REFERENCES

1 P. FAUVEL, *Faune de France*, Vol. 5, *Polychètes errantes*, Lechevalier, Paris, 1923.
2 P. FAUVEL, *Faune de France*, Vol. 16, *Polychètes sédentaires*, Lechevalier, Paris, 1927.
3 D. A. DORSETT, *J. Marine Biol. Assoc. (U.K.)*, 41 (1961) 577.
4 R. D. BARNES, *Biol. Bull.*, 129 (1965) 217.
5 R. D. BARNES, *Biol. Bull.*, 127 (1964) 397.
6 W. C. MACINTOSH, *British Annelids*, Vol. 4, Part 1, *Polychaeta*, Ray Society, London, 1922.
7 J. VOVELLE, *Arch. Zool. Exptl. Gen.*, 95 (1958) 52.
8 J. VOVELLE, *Arch. Zool. Exptl. Gen.*, 106 (1965) 1–187.
9 E. A. T. NICOL, *Trans. Roy. Soc. (Edinburgh)*, 56 (1930–31) 537.
10 G. BOBIN, *Arch. Zool. Exptl. Gen.*, 80 (1939) 144.
11 P. S. PALLAS, *Miscellanea Zoologica*, The Hague, 1778, pp. 123–124.
12 W. C. MACINTOSH, *Ann. Nat. Hist.*, 13 (1894) 13.
13 P. FAUVEL, *Mem. Pontif. Acad. Rom. Nuov. Lincei*, 21 (1903) 14.
14 A. T. WATSON, *Trans. Liverpool Biol. Soc.*, 42 (1928) 1–60.
15 H. FAOUZI, *J. Marine Biol. Assoc. (U.K.)*, 17 (1931) 379.
16 J. HANSON, *Nature*, 161 (1948) 610.
17 R. H. HEDLEY, *Quart. J. Microscop. Sci.*, 97 (1956) 411.
18 J. VOVELLE, *Bull. Lab. Maritime Dinard*, 42 (1956) 10.
19 E. O. MUZII, *Publ. Staz. Zool. Napoli*, 36 (1968) 135.
20 E. W. BEHRENS, *Contrib. Marine Sci. (U.S.)*, 13 (1968) 21.
21 A. SOULIER, *Trav. Inst. Zool. Montpellier et St. mar. Cette*, 2 (1891) 53.
22 G. CASANOVA AND J. COULON-ROSO, *Compt. Rend.*, 264 (1967) 2049.
23 D. W. EWER AND J. HANSON, *J. Roy. Microscop. Soc.*, 65 (1945) 40.
24 J. VOVELLE, *Compt. Rend.*, 244 (1957) 2964.
25 J. VOVELLE, *Bull. Soc. Zool. France*, 81 (1957) 202.
26 J. VOVELLE, *Compt. Rend.*, 246 (1958) 472.
27 J. VOVELLE, *Proc. 16th Intern. Congr. Zool.*, *Washington*, 1 (1963) 99.
28 R. DEFRETIN, *Compt. Rend.*, 232 (1951) 888.
29 R. DEFRETIN, *Compt. Rend.*, 146 (1951) 91.
30 M. T. COURTOIS, *Rapport Inst. Biol. Marit. Wimereux*, not published, 1961.
31 R. DEFRETIN, *Compt. Rend.*, 132 (1939) 501.
32 R. DEFRETIN, *Compt. Rend.*, 135 (1941) 1258.
33 R. DEFRETIN, *Compt. Rend.*, 136 (1942) 87.
34 R. DEFRETIN, *Ann. Inst. Océan.*, 24 (1949) 117–257.
35 R. DEFRETIN, *Compt. Rend.*, 230 (1950) 2343.
36 R. DEFRETIN, *Compt. Rend.*, 143 (1949) 1208.
36a R. DEFRETIN, *Compt. Rend.*, 145 (1951) 115.
37 C. RIMINGTON, *Biochem. J.*, 34 (1940) 931.
38 J. TILLMANS AND K. PHILIPPI, *Biochem. Z.*, 215 (1929) 36.
39 R. BELCHER, A. J. NUTTEN AND C. M. SAMBROOK, *Analyst*, 79 (1954) 201.
40 L. A. ELSON AND W. T. MORGAN, *Biochem. J.*, 27 (1933) 1824.
41 Z. DISCHE, *Methods of Biochemical Analysis*, Part 2, Interscience, New York, 1955, p. 213.
42 I. WERNER AND L. ODIN, *Acta Soc. Med. Upsaliensis*, 57 (1952) 230.
43 R. DEFRETIN, *Compt. Rend.*, 145 (1951) 117.
44 C. NEUBERG AND W. CAHILL, *Enzymologia*, 1 (1936) 22.
45 S. M. PARTRIDGE, *Biochem. J.*, 42 (1948) 238 [Partition Chromatography, *Biochem. Soc. Symp.*, 1949, No. 3].

46 R. DEDONDER, *Compt. Rend.*, 230 (1950) 997.
47 R. JOHANSON, *Nature*, 172 (1953) 956.
48 M. G. M. PRYOR, P. S. BRUSSEL AND A. R. TODD, *Biochem. J.*, 40 (1940) 627.
49 R. H. HACKMANN, *Biochem. J.*, 54 (1953) 371.
50 J. D. SMYTH, *Quart. J. Microscop. Sci.*, 95 (1954) 139.
51 M. LAFON, *Arch. Zool. Exptl. Gen.*, 96 (1958) 90.
52 O. SCHMIEDEBERG, *Mitt. Zool. Stat. Neapel*, 3 (1882) 373.
53 G. N. GRAHAM, P. G. KELLY, F. G. E. PAUTARD AND R. WILSON, *Nature*, 206 (1965) 1256.

Chapter XIII

Constituents of Brachiopod Shells

MARGARET JOPE

Department of Geology, Queen's University, Belfast (Great Britain)

1. Introduction

Shell is a calcified structure consisting largely of inorganic material, calcium carbonate or phosphate. It has, however, grown as living tissue and retained a complex organic constitution, little studied until recent years. This organic part contains proteins which bear phylogenetic information in their constitution. They must contribute mechanical stability though any role they may play in calcification is still not clear. Some of the other constituents have taken part in shell growth; others may be incidental or play a more mechanical role.

(a) The nature of brachiopods

The brachiopods are marine invertebrates of great interest. They are a small compact group with an interesting taxonomy and long lineage. The phylum has unbroken ancestry as two distinct groups back to the early Cambrian and the survival of the genus *Lingula*, unchanged since the Ordovician, provides an early example of phosphatic calcification. This and their intricate microstructure makes them excellent subjects for biochemical and taxonomic study, although the amount of organic material obtainable from any one shell is small.

The brachiopods are bivalve structures, usually attached to the substratum by a stalk. They are filter-feeders, taking their food from sea water by sieving through the lophophore.

[749]

(b) Classification

The phylum Brachiopoda is divided into two classes, the Articulata and the Inarticulata, differing as much in the chemical composition of their shells as morphologically. Articulate shell contains about 99% inorganic material (largely calcium carbonate) and less than 1% organic material, mostly protein. Inarticulate shell contains calcium phosphate as the mineral phase in place of calcium carbonate, and much more organic material (25–55 %) as a chitin–protein structure[1]. The amount of organic material is highest in the Lingulacea making the shell horny and flexible. The small superfamilial group, Craniacea, is included taxonomically within the Inarticulata, though they have carbonate and not phosphatic shells and their shell proteins more resemble those of the articulates in amino acid composition, than the inarticulates; their shells, however, contain a little chitin[2].

2. Shell structure

(a) General structure

The bivalve shells open and close by muscular action along an articulated hinge line (Articulata) or with no hinge line (Inarticulata). The muscle fibres are attached to the inner surface of the shell by tonofibrils, creating a distinctive scar on the interior of the shell which persists even in fossils.

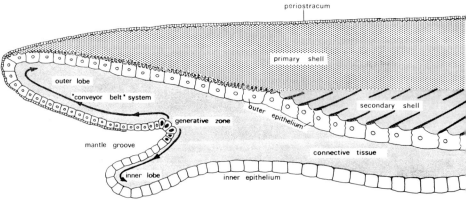

Fig. 1. Stylized longitudinal section of the edge of the valve of the articulate brachiopod, *Notosaria nigricans* showing periostracum, primary and secondary shell and their relationship to the outer mantle epithelium. (From Williams[6])

The brachiopod shell is a structure of great complexity, the details of which must be summarized. With articulates it is typically three-layered, consisting of (*i*) an outer covering, the periostracum, entirely organic, overlying (*ii*) an outer (primary) and (*iii*) an inner (secondary) calcareous layer (Fig. 1). Both calcareous layers contain a little organic material. With inarticulate brachiopods only the carbonate-shelled Craniacea have a true primary layer.

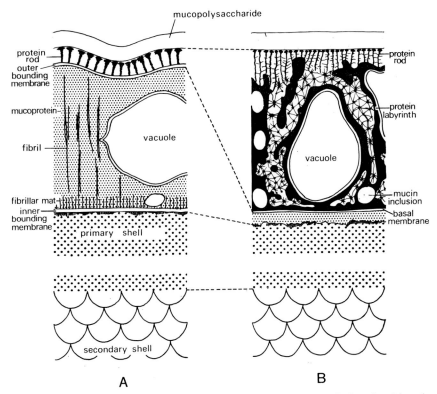

Fig. 2. Stylized sections of the periostracum and shell of the articulate brachiopods *Notosaria nigricans* (A) and *Waltonia inconspicua* (B) showing correlation of the two successions. (From Williams[7])

The details of the shell structure have been worked out by Williams and co-workers[3,4], most recently in electron micrographs which have revealed the highly organised fine structure and its relation to the epithelial cells of the mantle[5-11].

(b) Microstructures

(i) The periostracum[8,10,11]

The periostracum of both articulate and inarticulate brachiopods, as of some lamellibranchs, consists basically of a proteinaceous layer (on the order of 600 nm thick) between two membrane-like layers (12–14 nm thick) (Figs. 2, 3). With articulate periostraca, electron micrographs show the outer membrane has a triple-unit aspect and carries outwardly an array of erect fibrils. The fibrillar array is absent from the periostracum of the inarticulate Lingula[10] but seen in that of Crania where the outer membrane is single not triple-unit[11]. Yields of amino acids from brachiopod periostraca show these are mainly protein[12], present at least in part as protein–polysaccharide complexes, for acid mucopolysaccharides (metachromatic staining with toluidine blue) and neutral mucopolysaccharides (PAS reaction) have also been found in all the periostraca examined[13]. The bounding membranes contain lipids (see below).

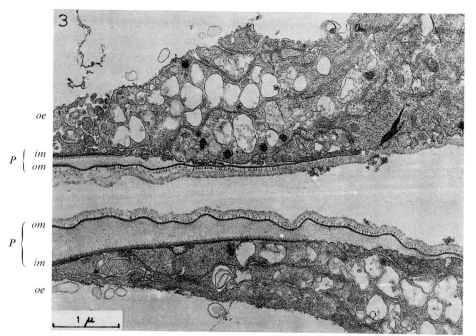

Fig. 3. Electron micrograph of a longitudinal section of the outer mantle lobe of a young articulate brachiopod, Notosaria nigricans, showing the range of layers contributing to the formation of the periostracum P, with its outer membrane om and inner membrane im, secreted by cells of the outer mantle epithelium oe. (From Williams[7])

The periostracum of phosphatic inarticulates (*Lingula* and *Discinisca*) contains chitin (identified by chitosan test and estimated as glucosamine)[1] apparently evenly distributed throughout. The periostracum of *Lingula* contains a significant quantity of iron (probably as hydrated carbonate)[1]; electron micrographs[10] of the mature periostracum show a narrow, outer, more electron-dense layer which may be tanned outer protein and a dark granular appearance which could be due to the iron.

The articulate periostracum shows more elaborate detail than that of inarticulates, and within the Articulata representatives of the orders Rhynchonellida and Terebratulida show a number of variants of the general pattern (Fig. 2).

The rhynchonellide periostracum[8]. The periostracum of *Notosaria nigricans* (Sowerby) consists of a relatively thick (600 nm) proteinaceous layer bounded by two 14-nm triple-unit membranes. The outer surface of the outer membrane carries an array of cylindrical units with expanded heads attached through pedestal-like extensions of membrane. These units, about 20 nm across and 35 nm high, are arranged in 75° rhombic array, 40 nm apart, bearing traces of 25-Å banding. The inner-bounding membrane shows a similar structure— a mat of erect fibrils 10 nm apart and about 30 nm high. The proteinaceous layer contains fibrils, possible extensions of the fibrillar mat; vacuoles bounded by 14-nm triple-unit membranes (fragments of outer-bounding membrane pinched off) and small mucin inclusions are also seen.

The periostracum of articulates may thus contain several protein molecular species: the protein of the membranes, the protein of the polysaccharide complex (by far the biggest contribution) and the fibrous protein of the fibrils and cylindrical units, which may be identical with or similar to protein of the polysaccharide complex. The pyrrolidine amino acid content and, in some genera, the glycine content of the periostracum[12] suggest at most only a small contribution from a collagen-type protein (Table I). The nature of the protein–polysaccharide complex requires further chemical identification.

Repeating structures composed of protein, attached to the surface of a membrane are not uncommon[14,15]. These usually appear to be enzymatic in function. Repeating particles are identified as terminal hydrolytic digestive enzymes, such as disaccharidases and peptidases[16], attached to the surface of intestinal microvilli membranes; and on mitochondrial inner membranes, head pieces (ATPases) attached by stalks to the inner surface of the membrane[17]. The primary nature of these particles is, however, questioned by Sjöstrand[18].

References p. 783

In brachiopods the periostracum outer- and inner-bounding membranes, about 140 Å thick, contain an electron-dense layer (about 40 Å) on either side, leaving an electron-transparent middle layer (about 50–60 Å), and though the layers are thicker, have the same triple-unit appearance as a plasma membrane (about 100 Å). These 140-Å triple-unit membranes have the same dimensions and appearance as the organic sheets within the calcareous secondary layer where they have been shown to contain protein, lipid and polysaccharide[19]. Their contribution to the total content of the periostracum is small, but to the organic material of the calcareous secondary layer it is much greater; their composition is discussed below.

(ii) The calcareous shell

Although the basic structure of the periostracum is similar throughout all groups of brachiopods, the structure of the calcified shell varies considerably between articulates and inarticulates. Even the carbonate-shelled inarticulates (Craniacea) with their division into primary and secondary calcareous layers have completely different organisation of mineral and organic constituents from either the articulates or phosphatic inarticulates.

(A) Articulata

(1) The calcareous primary layer[3–6] is made up of very fine fibrous calcite with the crystal *c*-axes normal to the shell surface. The thickness of this layer is uniform throughout the shell since it depends on the number of concentric rows of epithelial cells involved in its secretion (Fig. 1), fairly constant in adult shells — about 15 in *Terebratulina*. The mass of fibrous calcite contains protein material as irregular trails more or less normal to the secreting surface. These trails could be either microvilli of the mantle-epithelial cells, caught up in the calcite deposition (and should then contain also lipid and polysaccharide), or merely small amounts of protein exudate. The amino acid composition of the protein of this layer is given in Table II.

(2) The calcareous secondary layer[3–6]

(α) Terebratulida. The calcareous secondary layer of *Terebratalia transversa* (Sowerby) contains a series of long thin calcite fibres inclined at 10° to the primary layer, each surrounded by an organic sheet (appearing triple-layered in electron micrographs). On decalcification these sheets cohere to give a thick organic mat, similar in shape to the calcified shell and showing all its superficial details. The fibres have a distinctive cross section, related to the attitude of the secreting surface of the parent cells responsible for their growth, and are stacked in a series reflecting the or-

derly arrangement of these cells (Fig. 4A). Unlike the primary layer which does not increase in thickness during the life of the individual, the secondary layer thickens by continued deposition over the whole inner surface of the valve, as long as the cells remain functional, as well as at the margin of the valve where it represents the process of growth.

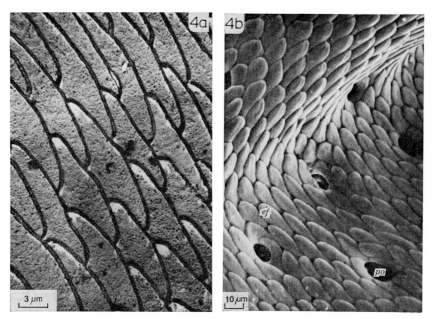

Fig. 4A. Electron micrograph of a transverse section of the secondary layer of the articulate brachiopod, *Macandrevia cranium*, showing stacking of fibres. (From Williams[6])

Fig. 4B. Scanning electron micrograph of the internal surface of the socket of the brachial valve of an articulate brachiopod, *Waltonia inconspicua*, showing the distribution of internal openings (po) of punctae which accommodate caeca, in relation to the exposed faces of secondary fibres (ef) — shell mosaic. (From Owen and Williams[9])

On the inner surface of this inner calcareous layer, the end-faces of the fibres and the organic sheets around them form a characteristic pattern — the shell mosaic (Fig. 4B), an imprint of the outer surface of the epithelial cells secreting this layer. The mosaic elements increase in size away from the margins of the valve, reflecting the increase in size of the epithelial cells with growth of the brachiopod. The mosaic pattern tends to be blurred at muscle emplacements. As these migrate and expand during growth there is

References p. 783

breakdown of the organic sheets separating the fibres giving irregularly sutured surfaces which become overlaid by further deposition of calcite. Similar breakdown of organic sheets resulting in blurred outlines to fibres occurs where secondary features such as teeth and crura are developed.

Within some species the mosaic pattern is very consistent, a typical species characteristic; in others the pattern is much more variable.

(β) *Rhynchonellida*. The rhynchonellide shell shows a modification in the disposition of the calcite fibres as compared with the terebratulide. The calcite fibres separated by triple-unit organic sheets are arranged in a series of layers (*e.g.* 6 in some individuals), one fibre thick with long axes subparallel to the surface of the valve. On dissolving out the calcite the triple-unit sheets cohere to a thin birefringent layer, appearing striated under the light microscope due to alignment of the sheets[13]. Intercalated between these layers of calcite fibres are organic sheets (protein) showing no substructure in electron micrographs[20], thicker and darker than the triple-unit sheets, especially in *Hemithiris*, less so in *Notosaria*. Under the light microscope they have a somewhat reticulated appearance, and are not birefringent[13].

These triple-unit sheets separating calcite fibres have a regular 140-Å thickness with a 50–60-Å electron-transparent middle layer (seeming to represent a considerable degree of organisation), of the same dimensions and substructure as outer- and inner-bounding membranes of periostracum, and with the same substructural appearance as the thinner (100 Å) mantle-cell plasma membrane. They grow as arcuate sheets attached to the anterior face of a mantle cell, but it is difficult to interpret these triple-unit structures as biologically functional membranes. It is necessary to know their true function and their molecular structure. It is generally agreed that not all membranes have a simple lipid-bilayer structure, as forms the basis of plasma membranes[18,21–26]; there is a range of types extending to the pure protein of bacteriophage head membranes and gas vacuole membranes of the blue-green alga *Microcystis aeruginosa*[26a]. There is also a range of thicknesses up to the cell membrane of the amoeba *Hyalodiscus simplex* with 3 layers each 44 Å thick[27]. In the brachiopod 140-Å triple-unit sheet the electron-transparent middle layer is thicker (50–60 Å) than two usual phospholipid–hydrocarbon chains (including their polar heads); the phosphorus-to-phosphorus distance across a synthetic bimolecular lipid leaflet is 51.5 Å (constant through a wide range of chain lengths[21]), indicating a structure not exactly analogous to the plasma membrane. These brachiopod shell triple-unit sheets seem to contain at least 10% lipid[19] (see p. 776), but considerably less than is normal for plasma membranes of red cell ghosts (20–40%)[26,28]. The triple-unit appearance may not reflect the native state of the brachiopod shell sheets. The ease with which phospholipid micelles may form *in vitro*, and can adsorb protein on their surfaces[22], is one way that a lipid bilayer might form spontaneously if lipid–protein interrelations were disrupted by denaturation of protein, to give a triple-layered electron micrograph substructure.

The corresponding picture in *Crania* is an unlaminated layer (about 130 Å thick)[11] largely protein, suggesting that with articulates the interior staining may be suppressed, perhaps due to the layout of particular polar and nonpolar side-chains, caused by folding in the tertiary molecular configuration.

(*3*) *Caeca*[9]

In brachiopods with punctate shells (among living forms, the Terebratulida and the Craniacea) the calcareous shell is penetrated by small canals (*e.g.* up to 20 μm across in *Macandrevia*) running from the mantle through the secondary layer, in a direction normal to the primary layer in which they

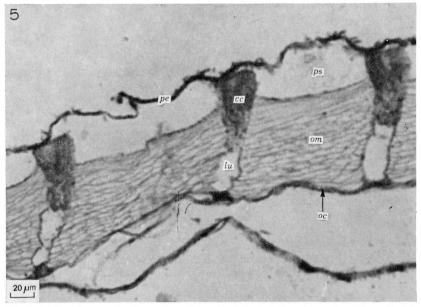

Fig. 5. Light micrograph of a section of the decalcified shell and mantle of the articulate brachiopod, *Macandrevia cranium*, showing the disposition of caeca with identifiable core cells (cc) and lumina (lu) in relation to the periostracum (pe), the space occupied by the primary shell (ps), the organic membranes (om) of the secondary shell and the outer epithelium of the mantle (oc). (From Owen and Williams[9])

end, just beneath the periostracum. The canals contain multicellular outgrowths of the mantle — caeca, originating in its outer lobe (Fig. 5). The part of the caecum which runs through the secondary layer is a stalk built up of flat peripheral cells surrounding a central lumen. At the junction to the primary layer the stalk expands to a head where the flat peripheral cells enclose core cells. The head is bounded by an organic membrane (45 Å thick, probably protein) continuous with the organic sheets covering the calcite fibres. A canopy of calcite (1 μm thick) overlying the protein membrane is densely perforated by canals (0.2 μm diameter) containing hollow extensions of the membrane which join the terminal cells of the caecum to the base of the periostracum. Breaks (up to 50 nm diameter) in the basal layer of the periostracum may connect the caecum with the outer layer of the periostracum.

In the caeca components such as glycogen, protein–polysaccharide complexes and small amounts of lipid are stored (and possibly synthesized) within the core and peripheral cells, and from here they are released into the mantle. Secretion of mucopolysaccharides from the caeca onto the surface of the shell may discourage attack from microorganisms.

Contribution of the caeca to the organic components of the calcareous layers. Since caeca are an intimate part of the organic structure of the calcified shell (in Terebratulida and Craniacea), their contribution to the organic components of the calcified shell must be allowed for when punctate and impunctate forms are being compared. Decalcification of the shell (5 % EDTA) and subsequent washing of the organic web allows some of the contents of the caeca to be washed away. The major fraction remaining consists of cell wall material, contributing, therefore, components of the mantle epithelium: protein[12] (about 70%) (Table III), lipids (about 20% — cholesterol and phospholipids, discussed below), small quantities of acid and neutral mucopolysaccharide and bound hexoses (5 %)[19]. Disc electrophoresis has not however shown multi-component protein patterns[12].

(*B*) *Inarticulata*

No three-layered structure comparable with that of articulate shells is seen in any inarticulate shell. The superfamilies Lingulacea, Discinacea and Craniacea each have a different calcareous shell structure.

(*1*) *Phosphatic shells*

(*α*) *Lingulacea.* The calcareous shell of *Lingula anatina* Lamarck consists of a series of lamellae (on the order of 12) subparallel to the surface of the valve, all containing chitin and protein but with varying mineralization; individual lamellae vary irregularly in thickness. Heavily mineralized lamellae tend to alternate with less mineralized ones; a few lamellae appear to be almost unmineralized, notably the outermost one which invariably carries a blue-green pigment; other apparently unmineralized lamellae carrying this pigment are seen sporadically throughout the shell and lenses of little-mineralized lamellae can persist in heavily mineralized ones. Mineralization is thicker in that part of the shell lying over the body cavity where some lamellae appear almost entirely mineral. Mineralized lamellae may contain regular trails of material normal to the shell surface — presumably calcium phosphate since they disappear on treatment with EDTA[13].

The organic material throughout the shell has a fine structure seen under the light microscope as alternating dark and light fine bands parallel to the shell surface. In electron micrographs[10] these are seen to be dark and light bands, 2.5 μm thick, suggesting layers predominantly protein alternating with layers predominantly chitin.

(β) *Discinacea*. In the calcified shell of *Discinisca lamellosa* Brøderip, the inorganic and organic constituents are more uniformly distributed than in the shell of *Lingula*. The valves are made up of a large number of lamellae composed of calcium phosphate within a chitin–protein ground mass. The outermost lamellae are oblique to the shell surface; underneath these the lamellae are subparallel to the inner surface of the valve giving an unconformity between the layers, which appears in transverse section as a division into outer and inner calcareous layers. The shell is traversed by fine punctae which branch towards the inside of the valve.

The organic material throughout the shell has a fine structure seen under the light microscope as alternating dark and light fine bands parallel to the shell surface (like those seen in *Lingula*)[13], presumably corresponding to the 2.5-μm layers seen in electron micrographs of *Lingula*, suggesting highly proteinaceous layers alternating with highly chitinous layers.

(2) Carbonate shells: Craniacea

The calcareous shell of *Crania anomala* (Müller), while resembling the carbonate shells of articulates in having a primary and a secondary layer and in containing true caeca, differs from them very much in the microstructure of these layers.

(α) *The calcareous primary layer*[11]. The mineral part consists of acicular crystallites (150 nm thick) inclined at 45° to growth surfaces within the layer, their rhombohedral cleavages giving the appearance, in longitudinal section, of a bedded succession. Traces of a growth lineation, with periodicity of about 200 nm, can be seen along the crystallites. These appear to be surrounded by organic material only towards the outer surface of the layer.

(β) *The calcareous secondary layer*[11]. The calcite of this layer is laminated. Laminae 260–320 nm thick are embedded in proteinaceous material in layers varying from 10 nm where laminae are thickest to 300 nm where the laminae are thin. The laminae rarely begin as perfect tabular rhombohedral seeds which grow evenly in all directions; usually they nucleate along linear changes in level of the protein foundation and then grow spirally from single or double screw dislocations. Extra hexagonal or dihexagonal edges are built into the crystallite boundaries and along with the spiral growth give a rounded outline to the crystallites, and crystallites amalgamate to give, with the addition of minor superficial spiral growth, a composite plate 200–250 nm thick.

The organic material of this layer[2] consists very largely of protein; some chitin is present (Table II). The decalcified organic web gives a positive PAS reaction indicating neutral mucopolysaccharides, but a negative reaction with toluidine blue indicating absence of acid mucopolysaccharides. The caeca, however, stain strongly with both these reagents[13].

(γ) *The muscle emplacements*. In the decalcified shell of *Crania* the *muscle scars* are clearly distinguishable from the remainder of the protein web as stiff white opaque chitinous deposits: in particular the four adductor muscle scars which radiate out from the apex of the valve tracing out the path of the muscle emplacement during growth. The presence of chitin in the scar has been shown by a positive chitosan test on the dissected out scars and a quantitative estimation from the glucosamine released on acid hydrolysis[2]. The chitinous pads survive heating in saturated KOH at 160° as small perforated discs about 1 mm across. The chitinous deposit, which may be a thin layer or a comparatively thick pad, is traversed by caeca whose contents stain strongly with both toluidine blue and PAS reagent, indicating the presence of acid and neutral mucopolysaccharides; the proteinaceous material immediately surrounding the chitin also stains strongly with both these reagents. The chitinous pad itself does not stain with PAS reagent in contrast to the strong reaction of the caecal contents. The non-chitinous regions of the protein web stain with PAS reagent but not with toluidine blue indicating neutral polysaccharides probably covalently linked to the protein[13].

Emplacement of the muscles causes even more alteration in the microstructure of the shell of *Crania* than in articulate shells. Electron micrographs[11] show that the regions of the muscle scars are thickened by a wedge of non-laminar calcite: crystallites deposited in a series of blades normal to the growth surface. The surfaces of the muscle scars are differentiated into polygonal areas and their wide boundaries (reflecting the outlines of outer epithelial cells and their intercellular spaces); the areas contain small rhombohedral units of calcite, and the wide boundaries, closely stacked platelets. The thickening of the muscle scar is associated with exudation of a dense mass of protein around the calcite followed by a 5-nm layer of protein attached to the mass. Chitinous material, appearing fibrous, is secreted on to the protein layer, followed by a thin layer of protein over the ventral surface of the chitin. Tonofibrils just penetrate the surface of the chitin, being held by fine fibrils which thus apparently form the muscle junction with its emplacement.

3. Shell constituents

(a) Inorganic

The inorganic elements in brachiopod shell[1,29] may be discussed in three groups:

(a) Those of major structural constituents, chiefly calcium — as calcium carbonate in articulate shells (94.6–98.6% of the inorganic material) and Crania (87.8–88.6%) and as calcium phosphate in inarticulates (lingulids 74.7–93.7%, Discinisca 75.2%). Smaller amounts of magnesium, as carbonate, are almost always found (articulates 0.4–1.4%, Crania 8.6%, lingulids 0.6–3.0%, Discinisca 6.7%). The processes of shell calcification are discussed below (p. 778).

(b) Lesser components, but nevertheless essential for enzyme and other biological functions. The latter may not all necessarily proceed in the shell itself, but the presence of such elements may provide a readily available reservoir for the brachiopod generally. Fe is found in small amounts in all forms; the periostracum of Lingula contains a relatively high proportion of Fe in some areas[1] (seen in electron micrographs[10] as small granules) which could be hydrated carbonate or possibly conjugated with an organic residue (the coelomic fluid of Lingula contains hemerythrin[30]). Na is found in small amounts in Terebratulina[31]; Mn and Cu, in traces in Terebratulina[31] and Lingula[29].

(c) Lesser components and trace elements with no known biological function, present purely incidentally, but not harmful: Al, Si and S (as sulphate) are found in all forms in small amounts; in trace amounts: Sr in *Macandrevia*[32], *Hemithiris*[33], *Terebratulina*[31] and *Lingula*[29]; Ba in *Terebratulina*[31] and *Lingula*[29]; B in *Terebratula vitrea*[34], *Terebratulina*[31] and *Lingula*[29]; Ti and Zr in *Lingula*[29]; Ta in *Terebratulina*[31]; F, 1.6–1.9% in *Lingula* as carbonate fluorapatite (francolite)[35], as a means of eliminating it from the system, or it might have a function analogous to that in dental enamel[36].

Recent spectrographic analysis[31] on valves of *Terebratulina retusa* Linné have shown the presence of Na as minor constituent, and of Fe, Cu, Mn as trace constituents in group (*b*); Al and Si as minor constituents and Sr, Ba, B as trace constituents in group (*c*). No significant differences in inorganic composition between valves of *Terebratulina* taken from different localities (Crinan Loch, Argyll, Scotland; coastal waters off Norway; coastal waters off Plymouth) were observed either in emission spectrograms or in γ-spectra (after neutron activation) indicating that the presence of these elements does not depend on the environment.

The following elements though specifically sought in *Terebratulina* were not detected — F, Co, Ni, Pb, K, Zr, Va, Sb, Zn, Mo, Ti, U, As. It is significant that F, Co, Zn, Mo were not detected even though essential for some enzymes.

No significant amounts of radioactive elements were detectable. This point was examined particularly to see if radiation damage might account for appreciable and anomalous disruptions of protein and other organic molecules[37].

(b) Organic

(i) Proteins

(1) Introductory

Brachiopod shell, like most calcareous living tissue, can be decalcified to leave a web of organic material, mainly protein (and with inarticulates, chitin). These structural proteins are intractable, resisting solubility in aqueous media, and have been little studied until recently. Their ranges of amino acid compositions are now becoming known[2,12], but their sequencing has so far not been much studied. They give, as the keratins with vertebrates, useful examples of intracellularly produced structural proteins from among the invertebrates.

Their biological context implies that these structural proteins should resist solubility in aqueous media. Yet they are not unduly low in side-chain polar groups, and their insolubility in their natural state must be due to the masking of these, probably concealed within the folding of the molecular tertiary configuration. They can be taken slowly into aqueous solutions, usually requiring urea[12], but this will have deranged any folded tertiary configuration of their natural state. They can sometimes be taken into various non-polar solvents (*e.g.* formic or dichloracetic acids)[12], and for future study of these proteins in their fully native configuration such solvents might perhaps be more exploited (for instance for optical rotatory dispersion), along with molecular studies by X-ray diffraction and electron microscopy on whole fibres.

In aqueous solutions these proteins seem through disc electrophoretic studies[12] to be long polypeptide chains, of molecular weight order 100000 (by comparison with other molecules, *e.g.* collagens). These units do not break up reversibly and must represent monomer molecules; they have a limited tendency to form polymers in solution. These structural proteins may, however, be composite.

The data so far available reveal no high degree of ordered molecular arrangement for these shell proteins, such as gives the 640-Å banding of collagens. By their insolubility, these molecules are hardly in the simple state of long chains when in their native position in the shell; the masking of lyophilic groups is perhaps not even by regular coiling — the molecule may be arranged in a more random meshwork. Though no strong interchain linkages (*e.g.* S–S) need be broken to carry them into aqueous solution[12], meshworks comparable with elastin or particularly the rubber-like isotropic resilin[38] might be held by hydrogen bonding.

The brachiopod structural proteins preserve some fairly closely defined properties through their considerable range of amino acid composition[2,12]. The variations in composition reflect to a fair extent the zoologically reasoned taxonomy within the phylum, and must largely represent cumulative non-lethal mutations in the protein biosynthesis instructions. The brachiopod amino acid data have thus demonstrated the basis for a structural protein taxonomy well related to the zoological; it extends to families and in some cases to genera (*e.g.* the high tyrosine throughout the structural proteins of *Laqueus*), and stresses the anomalous position of *Crania*.

Some of the properties of these structural proteins may be due partly to their amino acid constitution. Others, notably their resistance to solubility, must arise (perhaps through folding) during biosynthetic assemblage of the molecule[39], presumably on the polysomes of the mantle epithelial cells. The change of composition (and perhaps character) between periostracum protein and the protein of the calcareous shell itself[2,12], produced by the same cells

TABLE I

AMINO ACID COMPOSITIONS OF PROTEINS FROM BRACHIOPOD PERIOSTRACUM

Amino acids are stated as residues per 100 total amino acid residues

From Jope[12].

Phylum	Brachiopoda										Inarticulata	
Class	Articulata										Inarticulata	
Order	Terebratulida										Acrotretida	Lingulida
Genus	Laqueus	Laqueus	Neo-thyris	Walt-onia	Terebratalia	Terebratalia	Macandrevia	Macandrevia	Gry-phus	Tere-brat-ulina	Crania	Lingula
Protein source					Periostracum							
Gly	32.4	34.9	39.2	28.8	21.4	22.6	22.6	30.4	21.9	15.2	14.0	22.8
Ala	3.7	2.4	4.2	7.0	6.0	3.9	5.0	3.7	6.4	8.8	9.4	11.6
Val	2.6	3.2	3.2	4.5	6.2	5.0	3.6	3.3	5.2	6.5	7.4	4.4
Leu	2.0	2.3	3.0	4.5	4.4	4.2	3.6	3.1	4.2	6.5	7.4	3.4
Ile	1.2	1.6	2.3	3.8	3.9	3.6	3.2	2.7	4.0	5.1	5.1	2.3
Ser	7.5	4.9	8.5	7.1	4.3	5.5	9.9	6.6	9.3	6.7	8.6	6.6
Thr	1.6	2.2	1.5	3.3	3.9	2.4	3.2	2.3	4.6	5.2	6.3	4.5
Pro	5.2	5.0	1.9	3.0	4.5	4.4	3.7	3.1	3.6	3.8	2.3	7.6
Hyp	—	Tr	—	Tr	0.5	1.3	0.5	1.2	—	—	—	2.9
Phe	6.1	6.1	3.1	4.4	4.1	5.0	2.8	3.1	3.4	4.1	3.8	3.7
Tyr	5.9	4.8	0.3	—	4.1	6.1	3.8	3.2	1.2	2.0	1.1	3.5
Cys	1.3	1.4	0.6	1.1	5.6	4.6	2.2	1.7	1.4	2.0	Tr	2.8
Met	0.1	0.2	Tr	Tr	0.6	0.7	0.8	0.4	Tr	1.0	1.0	0.9
Asp	13.5	11.3	15.2	14.6	11.7	10.4	13.4	12.7	11.8	11.3	12.4	8.7
Glu	4.7	5.4	6.6	7.5	5.9	5.1	6.0	4.7	7.3	9.1	9.2	4.7
Hyl	—	—	—	—	—	0.3	—	0.4	—	—	—	—
Lys	2.8	3.0	3.4	3.7	4.6	5.6	5.4	4.1	4.2	4.1	5.0	2.7
His	3.2	4.1	2.0	1.1	0.4	1.5	3.9	6.1	3.1	1.6	1.9	0.9
Arg	6.4	6.1	5.3	5.3	8.9	8.2	6.0	5.6	8.5	5.7	5.0	6.3
Pro+Hyp	5.2	5.0	1.9	3.0	5.0	5.7	4.2	4.3	3.6	3.8	2.3	10.5
Gly+Ala	36.1	37.3	43.4	35.8	27.4	26.5	27.6	34.1	28.3	24.0	23.4	34.4
Gly/Ala	8.8	14.5	9.3	4.1	3.6	5.7	4.5	8.1	3.4	1.7	1.4	2.0

a little later on in their life history, may reflect a change in determinants of this biosynthesis.

It is still not clear what role these structural proteins of shell may play as determinants in the processes of biological calcification.

(2) *Component protein molecular species*

The *periostracum*, the non-calcified outer layer, contains protein as part of a membrane-like structure[8] as well as protein of a polysaccharide complex (the main bulk of the periostracum). Amino acid data for whole periostracum varies more between individuals than that for calcareous shell proteins[2,12] — due partly to erosion or contamination of its surface (an external cover) by microorganisms or other causes, but possibly also to variations in proportions of different proteins in individuals (Tables I and II).

The protein of this layer is more difficult to take fully into solution than the calcareous shell proteins, due perhaps to its quinone tanning; it has not yet been fractionated.

The *calcareous shell proteins*[2]. The organic material remaining after decalcification of the shell consists of a number of protein-containing components, varying in different groups of brachiopods:

Articulata Terebratulides, triple-unit layers + caeca (mainly epithelial cell wall protein from the mantle) (Fig. 5).
Rhynchonellides, triple-unit layers + single-unit layers.
Inarticulata Craniacea, single-unit layers + caeca.
Discinacea, layers of protein–chitin complex.
Lingulacea, layers of protein–chitin complex.

The calcareous shell proteins (all groups, except rhynchonellides, owing to insufficient material) on *disc electrophoresis* migrate as a single homogeneous fraction[12]. A mixture of proteins from different genera, showing very different amino acid compositions, however, also migrates as a single fraction, indicating that the components will be of closely related molecular dimensions and behaviour. Mantle protein contains a small proportion of a slower component, interpreted as polymer which was, however, not detected in any of the calcareous shell proteins. The main fraction of the mantle proteins migrates as one fraction with the calcareous shell proteins. Molluscan extrapallial fluid examined by paper and by cellulose acetate electrophoresis shows a single protein fraction when the shell is calcite alone, but three or more when it is aragonitic[40].

TAB

AMINO ACID AND AMINO SUGAR COMPOSITIONS OF PROTE

Amino acids are stated as resi

F

Phylum												Brachiop
Class				Articulata								
Order				Terebratulida								
Genus	Laqueus					Neothyris			Waltonia[a]	Tere-bratalia[b]	Macan-drevia	Terebratulin
Protein source in shell	Peri-ostra-cum	Outer and inner calcite			Inner cal-cite	Peri-ostra-cum	O and I calc.	O calc.	I calc.	O and I calc.	O and I calc.	O and I calc.	O an calc
Gly	32.4	20.9	18.2	19.9	28.8	39.2	26.8	24.7	31.7	22.6	23.4	19.6	25.8
Ala	3.7	4.9	5.4	4.4	5.5	4.2	12.6	11.7	14.3	11.0	7.4	7.1	5.1
Val	2.6	6.9	6.5	6.5	7.0	3.2	5.6	6.0	4.2	6.0	7.8	7.8	9.9
Leu	2.0	4.3	5.2	4.6	3.4	3.0	4.2	5.0	3.0	4.8	4.3	5.6	5.2
Ile	1.2	3.8	4.6	4.5	3.6	2.3	3.8	4.2	2.6	4.7	3.7	4.3	5.5
Ser	7.5	8.5	7.5	8.1	6.0	8.5	6.6	7.1	6.4	7.0	9.0	9.1	8.6
Thr	1.6	4.8	5.1	5.1	4.2	1.5	3.7	3.9	2.7	4.0	5.1	5.3	4.8
Pro	5.2	4.0	5.8	5.1	2.3	1.9	5.1	5.0	4.6	7.3	5.8	6.0	2.9
Hyp	—	—	—	—	—	—	—	—	—	—	—	—	—
Phe	6.1	4.0	3.4	3.9	3.4	3.1	3.5	4.0	3.0	3.5	3.4	3.0	2.4
Tyr	5.9	6.2	5.3	6.3	6.8	0.3	—	—	0.1	—	—	—	—
Trp	c	c	c	c	c	c	c	c	c	c	c	c	c
Cys	1.3	3.3	3.7	5.3	3.9	0.6	2.4	1.1	4.7	1.2	Tr	—	0.1
Met	0.1	1.6	1.7	1.8	1.0	Tr	1.3	1.3	0.9	1.1	1.4	1.6	1.1
Asp	13.5	10.1	9.2	8.2	9.2	15.2	10.0	10.1	9.6	10.5	10.5	12.3	12.1
Glu	4.7	6.9	7.2	6.1	5.6	6.6	5.7	6.5	4.6	6.4	6.5	8.6	7.9
Hyl	—	—	—	—	—	—	—	—	—	—	—	—	—
Lys	2.8	3.2	4.0	3.6	2.7	3.4	2.6	3.2	2.0	3.8	4.1	3.9	3.7
His	3.2	1.0	0.8	0.7	0.5	2.0	0.1	0.2	0.2	0.3	0.6	3.0	0.6
Arg	6.4	5.5	6.3	5.9	5.8	5.3	6.1	6.0	5.6	5.8	7.1	5.7	8.3
Glucos-amine	—	—	—	—	—	—	—	—	—	—	—	—	—
Gly+Ala	36.1	25.9	23.6	24.3	34.3	43.4	39.4	36.4	46.0	33.6	30.8	26.7	30.9
Gly/Ala	8.8	4.2	3.3	4.5	5.2	9.3	2.1	2.1	2.2	2.1	3.2	2.7	5.0
Dicarb/ Basic	1.5	1.8	1.5	1.4	1.6	2.0	1.8	1.8	1.8	1.7	1.5	1.7	1.6

a Previously designated Terebratella.
b Without periostracum.
c Not estimated.

M THE VARIOUS LAYERS OF BRACHIOPOD SHELLS

00 total amino acid residues

₂.

		Rhynchonellida		Inarticulata							
				Acrotretida			Lingulida				
	Hemithiris	Notosaria		Crania		Discina	Lingula				
…ft …ein …er	Hard protein layer	Whole calcite protein		Whole calcite protein		Whole shell[b]	Outer layer	Mid layer	Inner layer		
,5	26.0	33.2	35.1	14.1	14.4	22.2	15.3	15.6	13.0	13.7	Gly
8	5.2	8.6	9.6	6.1	7.1	18.8	21.5	15.6	16.6	17.9	Ala
,2	7.1	3.1	3.3	5.0	4.6	1.9	5.3	3.7	7.1	6.0	Val
,6	5.8	2.3	2.5	4.5	4.6	1.5	3.5	4.8	4.6	3.7	Leu
,1	3.7	1.4	1.5	3.0	3.8	1.0	2.1	2.8	2.8	2.1	Ile
,6	6.3	6.9	6.5	14.0	12.9	3.5	5.3	6.3	5.6	4.9	Ser
2	3.7	2.7	2.7	5.8	5.3	2.7	3.4	4.0	3.9	3.2	Thr
1	6.0	8.8	7.2	4.9	6.6	5.5	6.7	7.1	7.7	7.7	Pro
	—	—	—	—	—	13.7	4.3	6.3	5.5	6.1	Hyp
2	2.5	8.3	6.2	2.2	2.9	0.7	2.1	2.3	2.6	1.8	Phe
	—	0.2	—	—	0.2	0.1	2.5	2.2	2.7	1.8	Tyr
c	c	c	c	0.1	c	c	0.5	c	c	c	Trp
,6	0.7	0.2	1.3	1.5	Tr	0.7	0.1	0.6	0.3	0.1	Cys
,3	0.7	0.9	0.7	1.0	0.9	0.1	0.5	0.5	0.9	3.7	Met
,5	10.6	11.9	10.6	18.5	18.4	7.0	10.9	11.3	10.3	9.9	Asp
,2	5.2	4.6	4.4	8.8	9.4	4.9	5.8	6.1	5.9	5.2	Glu
	—	—	—	—	—	0.1	—	0.1	—	—	Hyl
,4	2.7	1.3	1.6	4.5	4.8	1.9	2.0	2.5	2.7	5.0	Lys
,4	2.8	1.5	0.9	1.9	1.7	4.5	0.8	0.9	0.9	0.9	His
,3	11.6	3.9	4.0	4.2	5.2	8.3	7.6	7.0	6.8	6.1	Arg
	—	—	—	3.4	5.4	1.7	5.6	3.8	7.4	6.3	Glucosamine
,2	31.2	41.7	44.7	20.2	21.5	41.0	36.8	31.2	29.6	31.6	Gly+Ala
,1	5.0	3.9	3.7	2.3	2.0	1.2	0.7	1.0	0.8	0.8	Gly/Ala
,5	0.9	2.5	2.3	2.6	2.4	0.8	1.6	1.3	1.6	1.7	Dicarb/Basic

TABLE III

AMINO ACID COMPOSITIONS OF PROTEINS FROM BRACHIOPOD MANTLES
Amino acids are stated as residues per 100 total amino acid residues.

From Jope[12].

Phylum				Brachiopoda						
Class		Articulata							Inarticulata	
Order		Terebratulida				Rhynchon-ellida		Acro-tretida	Lin-guli-da	
Genus	Terebrat-alia		Macan-drevia	Gry-phus	Terebrat-ulina		Notosaria		Crania	Lin-gula
Protein source					Mantle					
Gly	20.6	24.5	21.0	21.0	22.0	24.2	17.2	16.2	18.0	24.0	25.9
Ala	8.5	8.9	9.3	7.8	8.3	7.9	8.9	9.0	8.1	6.9	7.7
Val	4.5	3.2	4.9	4.3	4.6	2.8	5.7	5.4	5.4	2.2	1.8
Leu	5.1	4.1	5.1	5.1	5.4	4.6	6.3	6.4	6.4	3.8	3.4
Ile	3.6	1.8	3.7	3.8	3.6	2.2	4.3	4.2	4.3	1.9	1.2
Ser	2.8	5.3	3.4	4.3	3.6	6.2	3.6	2.6	2.8	3.5	4.3
Thr	3.6	4.2	3.6	4.2	3.5	4.4	4.2	3.6	3.6	2.9	2.5
Pro	10.1	8.1	4.4	8.6	6.0	8.3	2.3	8.6	6.6	12.2	12.3
Hyp	5.0	5.0	5.8	3.0	5.1	5.0	3.4	3.7	3.9	10.6	10.1
Phe	2.8	2.1	2.7	2.7	2.8	3.0	3.2	2.9	3.2	1.7	1.4
Tyr	1.9	2.0	2.5	2.1	1.9	1.8	2.8	2.5	2.4	0.9	0.7
Cys	2.2	1.5	1.9	2.1	1.6	—	1.4	1.5	2.2	0.6	0.2
Met	2.1	1.6	2.3	1.3	1.6	1.2	2.0	1.8	2.0	1.5	1.2
Asp	6.5	7.1	7.9	8.0	7.8	7.4	9.5	8.1	8.6	6.2	6.7
Glu	7.9	9.2	8.8	8.0	9.3	9.8	10.7	9.5	10.5	10.5	10.7
Hyl	0.9	1.0	1.3	1.3	1.1	1.0	1.0	1.6	0.9	0.1	1.2
Lys	4.1	3.6	3.5	3.6	3.8	3.5	5.7	5.1	4.5	3.1	3.3
His	1.0	0.9	1.2	0.9	1.3	1.0	1.5	1.4	1.4	0.4	0.4
Arg	6.1	6.2	6.6	6.9	6.3	5.9	6.1	5.9	6.3	5.7	4.9
Pro+Hyp	15.1	13.1	10.2	11.6	11.1	13.3	5.7	12.3	10.5	22.8	22.4
Gly+Ala	29.1	33.4	30.3	28.8	30.3	32.1	26.1	25.2	26.1	30.9	33.6
Gly/Ala	2.3	2.8	2.3	2.7	2.7	3.1	1.9	1.8	2.2	3.5	3.4
Dicarb/Basic	1.2	1.4	1.3	1.3	1.4	1.5	1.4	1.3	1.5	1.8	1.8

Solubility. The main bulk of the protein is very insoluble in mild aqueous media. Decalcification with 5% EDTA takes no more than a trace of protein into solution either from the periostracum or the calcareous shell. Acetic acid (7%) takes into solution a small proportion of the calcareous shell protein of *Terebratalia* (about 2%).

(*3*) *Chemical characteristics of the proteins*

(α) *Composition of the peptide chains: amino acid data*[2,12]

Articulata. Periostracum protein has generally very high glycine with low or moderate alanine (Table I). Aspartic acid is higher than in the calcareous shell protein and like the latter gives a greater number of acidic than basic groups. The toughness of the periostracum may be due to quinone tanning, either self-tanning using the tyrosine of the polypeptide chain or through introduction of a tanning agent such as hydroquinone. The proportion of tyrosine varies with genera, from very low or absent to moderately high. The periostracum of some genera, articulate as well as inarticulate, contains a little hydroxyproline, the former perhaps reflects the composite nature of this tissue. Periostracum, in general, tends to have fairly high histidine.

Calcareous shell proteins. The shells of the impunctate rhynchonellides contain two kinds of organic secretion, seen most clearly in *Hemithiris* (Table II): the "soft protein layers", where the protein is built into triple-unit layers, have very high glycine, high alanine and aspartic acid, and greater proportion of acidic than basic groups; the "hard protein layers" have high glycine, aspartic acid and arginine and greater proportion of basic to acidic groups.

The shells of the punctate terebratulides contain caeca (see above); since it has not been possible to separate these from the organic web, the amino acid data discussed here represent the protein of the web along with epithelial cell wall protein. The terebratulide proteins (Table II) have high glycine and aspartic acid and in some genera high alanine. Mantle proteins (Table III) contain hydroxyproline (about 5%), high glycine and high dibasic amino acids especially glutamic. The intrusion of mantle proteins into the terebratulide shell may be responsible for its higher glutamic acid content as compared with rhynchonellides. The absence of hydroxyproline in terebratulide shells indicates however that they contain no more than a small proportion of mantle protein or that the hydroxyproline-containing protein is not part of the caeca.

Inarticulata. Both the periostracum and calcareous shell proteins of the phosphatic inarticulates (Tables I and II) contain hydroxyproline, a higher proportion in the latter which also contain small amounts of hydroxylysine.

The calcareous shell proteins of this group have high alanine, *Lingula* being unusual with more alanine than glycine; alanine and glycine together make up about 30% of the residues. In *Discina* the glycine is only slightly higher than the alanine. The periostracum of *Lingula* has higher glycine than its calcareous shell (as usual for brachiopods) and fairly high alanine, although here the alanine is well below glycine. Aspartic acid is high in the calcareous shell protein of *Lingula* but less so in its periostracum and in the calcareous shell of *Discina*. The latter with its rather high arginine and histidine is an exception to the generalisation that brachiopod shell proteins have a greater number of acidic than basic groups.

The carbonate shelled inarticulate *Crania* like the terebratulides has caeca intruded through the calcareous layers and so the amino acid data for these also represent the protein of the organic web along with some epithelial cell wall protein. *Crania* has only moderately high glycine in both periostracum and calcareous shell protein. Aspartic acid is high in both but particularly so in the calcarous shell protein where it is higher than glycine. The calcareous shell has unusually high serine for brachiopod proteins (Tables I–III).

Hydrolysates of all the brachiopod shell proteins contain large amounts of ammonia, some of which may have come from dibasic amino acids, aspartic and glutamic, in the protein structure as amides.

General comparisons. Brachiopod shell proteins[2,12] generally resemble those of mollusca in having high glycine and sometimes high alanine, but the values for this large phylum (Mollusca) vary over a wide range[41]. Both phyla have particularly high glycine in the periostracum, usually higher than their calcareous shell proteins, but brachiopod glycine never reaches the 50% level found in some molluscan periostraca[42,43]. All brachiopods have high aspartic acid in both periostracum and calcified shell, higher in the periostracum except for *Crania* (18.5%). Molluscan aspartic acid is usually high in both periostracum and calcified shell but particularly in the latter where it may be as much as 32.7% (in the bivalve *Pitar morrhuana*)[44]. Like the molluscan, some brachiopod periostraca contain small amounts of hydroxyproline and hydroxylysine, which may suggest a composite protein.

(β) *Interchain linkages*[12]

Brachiopod proteins appear to have no interchain linkages: no ester-type or disulphide linkages can be detected even though cystine is almost always found in hydrolysates. The low tyrosine in some genera suggests the possibility of tyrosyl interchain cross-linkages (as in resilin[45]), yielding di- or tri-tyrosyl on hydrolysis of the peptide chain.

(γ) *Solubility*[12]

Brachiopod shell proteins in their native state are highly insoluble, but those of articulate shell can be taken into Tris–glycine buffer (pH 8.9) with the aid of 7 M urea, which presumably exposes lyophilic groups by loosening H-bonding and unfolding.

The chitin–protein structures of *Lingula* (after decalcification) appear unaffected by urea–Tris–glycine treatment; this material can be taken up in anhydrous formic acid from which it may be recovered after removal of formic acid (*in vacuo*); this residue may then be taken into Tris–glycine–urea buffer. The shell protein of *Crania* also dissolves in this buffer but the amount of chitin here is small (located at muscle emplacements).

(δ) *Tanning*

The toughness and yellow-brown colour of brachiopod periostracum suggests tanning, either self-tanning by tyrosine of the polypeptide chain or tanning by introduction of hydroquinone as in insect cuticles[46]. Tyrosine is very variable in brachiopod shell protein (but less variable in mantle) (Tables I–III) and often not found at all in calcified shell, where there is no appearance of tanning (see also p. 765).

(ε) *Ageing*

No differences have been observed in the solubility behaviour of brachiopod shell proteins as between younger and older individuals[13], such as should be observable with any progressive increase of cross-linkage with ageing (as with collagen-type proteins); such effects are found in postmortem changes.

(ζ) *Pigments*

Brachiopods do not show a generally developed pigmentation system in the shell. Their general yellow-brown colour is probably due to optical reflection and scattering effects and tanning in the periostracum. *Lingula* has a bright green colouring mainly in the outermost layer of the calcified shell, due to an unindentified pigment. It has also hemerythrin in its coelomic fluid[30]; otherwise brachiopods seem without oxygen-transport pigments in amounts observable to the eye.

(4) *Biochemical taxonomy in shell structural proteins*[2,12]

The amino acid data (Table II) on brachiopod calcareous shell proteins show some relation to the zoologically reasoned taxonomy, substantiating it and demonstrating that these structural proteins are direct products of genetically controlled biosynthetic systems. The data also confirm the intermediate and in some ways unique position of the small superfamilial group, the Craniacea.

References p. 783

First, the amino acid data distinguish the two classes within the phylum — Articulata with carbonate shell and Inarticulata with chitinous–phosphatic shell. The Articulata show high glycine but lower alanine, whereas the Inarticulata have less glycine but considerably more alanine, tending to equal the glycine. Further, only the Inarticulata yield glucosamine (from hydrolysis of chitin), the anomalous *Crania* having a little chitin located at muscle emplacements. The most striking distinction between classes is the hydroxy-proline-containing protein of inarticulate phosphatic shell, which is not found in the carbonate shell of articulates or in the carbonate-shelled *Crania*. A little hydroxylysine has been shown in phosphatic inarticulate shell. Hydroxyproline is, however, traceable in the periostracum of inarticulates and some articulates, though here again there is a distinction between the two classes, for the amount is considerably higher in the phosphatic inarticulate *Lingula*.

Secondly, distinctions between orders are shown up in ratios of dicarboxylic to basic amino acid residues. Within the Articulata, the Terebratulida have a lower ratio than the Rhynchonellida; and within the Inarticulata, the Lingulida (*Lingula*) differ from the Acrotretida, being higher than *Discina* (Discinacea) and lower than *Crania* (Craniacea).

Thirdly, some distinctions between families are seen: the Terebratellidae (*Neothyris* and *Waltonia*) are distinguished from other Articulata by their higher alanine values. At the genus level, *Laqueus* stands apart with high tyrosine.

The amino acid data for *Crania* reflect the anomalous taxonomic position of the Craniacea, classified zoologically as Inarticulata, but with carbonate shell. Its high glycine-to-alanine ratio and the absence of hydroxyproline classifies *Crania* biochemically with the Articulata but the presence of chitin in its shell stresses its relations with the Inarticulata. *Crania*, however, sets the Craniacea apart from both Articulata and Inarticulata by its high aspartic acid and serine.

The amino acid compositions of periostracum proteins show greater variability between individuals, and fewer taxonomic distinctions than the calcareous shell proteins. This tissue as an outer covering to the shell is subject to hardening, tanning and erosion by microorganisms or mechanically, conditions which may cause alterations in the polypeptide chain, or differential removal of parts of its complex structure, giving variations in amino acid content. The high pyrrolidine amino acid content of the phosphatic Inarticulata again distinguishes them from the Articulata and the carbonate *Crania*, as does their higher alanine and lower aspartic acid.

Thus the structural proteins of shell, as well as of mantle, are here shown to retain signs of their production by genetically controlled biosynthesis, though this may be overlaid by some post-biosynthesis effects, such as catabolic. There is also the question of the extent to which the location of proline residues becoming hydroxylated (hydroxylation takes place secondarily after assembly of the polypeptide chain[47,48]) is subject to genetic control. Those residues appearing as hydroxyproline in the final protein will all have been coded as proline in the initial peptide assembly, and this location could be dictated by the tertiary configuration of the molecule, a subsidiary outcome of the coded synthesis.

(ii) Polysaccharides

(1) Chitin

Chitin forms part of the exoskeleton of inarticulate brachiopods and as in the skeletal substance of many invertebrate groups is usually associated with protein, possibly a consequence of biosynthesis[49,50]. When in large proportion, it gives shells a horny appearance and texture in spite of its crystalline structure. The three crystal forms α-, β- and γ-chitin[51,52] correspond to long straight chains of β-1,4-N-acetylglucosamine, folded on themselves in α- and γ-chitin to give two and three parallel segments, respectively; one in six amino sugar residues seems to be deacetylated. The pedicle and setae of Lingula contain β-chitin[53] but the chitin of the shell has not yet been examined. All three forms can be found in the same individual (e.g. in Loligo[51]) and it is assumed that the manner in which the polysaccharide is synthesized and the nature of the protein to which it is attached determine whether the configuration will be α-, β- or γ-chitin. In common with other structural polysaccharides chitin has a β-glucoside link instead of the α-link usual in metabolic polysaccharides. Rudall has discussed the structure of α-, β- and γ-chitins in terms of X-ray and infrared data[51]. He concludes that β-chitin is reasonably seen as having one molecule of bound water per N-acetylglucosamine residue[54]. With α-chitin he emphasizes that X-ray diagrams and density correspond to anhydrous poly-N-acetylglucosamine, and concludes that deacetylation and spatial replacement by bound water probably makes only a small contribution.

Chitin has been shown in the periostracum and calcified shell of Lingula and estimated as glucosamine released on acid hydrolysis (up to 7% of shell organics[1,2]). Discina with lower organic content of the calcified shell gave only 1.7% glucosamine[2] (Table II). These low values for chitin in calcified shell suggest that only a proportion of the glucosamine is released by acid hydrolysis. Jeuniaux[55,56] has shown that enzymatic hydrolysis (chitinase and chitobiase) releases about 30% of the total chitin–free chitin; the remainder (masked chitin) can only be hydrolysed by these enzymes after destruction of material (chiefly protein) associated with the chitin probably as complexes. Alkaline hydrolysis (0.5 N NaOH at 100° for 6 h) in conjunction with enzymatic hydrolysis has yielded chitin (free+masked), 29% of the organic material in the calcareous shell of *Lingula*.

The periostracum of *Lingula* appears (in electron micrographs[10]) to have chitin and protein evenly distributed throughout, unlaminated. The mature periostracum has a thin outer layer of greater electron density, not seen in earlier stages of growth, which could be interpreted as tanning of the protein. Füller[57] considered the periostracum to have an outer layer of sclerotized chitin.

In the calcified shell of *Lingula* chitin and protein are distributed throughout the heavily mineralized as well as the unmineralized layers in laminae (2.5 μm in electron micrographs[10]), seen under the light microscope as alternating dark and light bands, fanning out gradually in places[13]. Similar laminae are known in structures such as arthropod procuticles (0.2–10 μm)[58] and insect cuticles[51,59] and are also seen to fan out; possible interpretations are that dark chitin–protein layers are separated by layers with less protein or that chitin–protein micelles are more tightly packed in the electron-denser laminae than in the less dense[58]. Within laminae the organisation may be of layers of protein and chitin fibres lying parallel to the surface of the laminae. X-Ray data[51] show three types of chitin–protein complex: (*i*) periodicities of 33 Å possibly corresponding to alternating layers of chitin and protein; (*ii*) a structure (possibly protein) repeating at 31 Å along the length of the chitin chain, could represent the attachment of a protein molecule to every sixth amino sugar residue and (*iii*) a structure repeating at every eighth amino sugar residue. A covalent link with protein has been shown, for α- and β-chitin, linking chitin to two histidine and one aspartic acid residue for each 400 acetylglucosamine residues[60] (but see also Attwood and Zola[61]). It could be the details of the chitin–protein association which cause a folding of the chains in groups of three (along the c-axis of the chitin). This may be brought about by the chain being fixed at its two ends to protein[51]. Rudall[62] considers that cuticular structures can contain, side by side, different chitin–protein complexes with one type predominating in a given material.

Chitin is presumably synthesized in the cells of the outer mantle epithelium.

Cells engaged in laying down the chitin are seen (in electron micrographs[10]) to be heavily charged with structures resembling glycogen bodies. These are a reserve of glucose and would be the ultimate source of the hexose unit, converted first to uridine diphosphoglucose then to uridine diphospho-*N*-acetylglucosamine which acts as the glycosyl donor[49]. Synthesis of chitin proceeds by glycosyl transfer from the nucleotide-linked sugar to preformed chitodextrin chains, catalyzed by polymerase enzymes.

Synthesis of chondroitin sulphate[50] seems to require prior synthesis of an acceptor protein; likewise, protein may be required before the repeating *N*-acetylglucosamine units can be assembled into the polysaccharide chain of chitin. Such an acceptor protein might explain why chitin seems never to be found in the free state in nature but invariably linked to protein.

(2) *Other polysaccharides*[13]

The organic web of the calcareous layers of *Notosaria* contains both acid and neutral mucopolysaccharides (about 2–5%) after proteolytic digestion, with about 5% hexose sugars. Mantle epithelium similarly contains acid and neutral mucopolysaccharides (about 5%), both extractable, in part, by hot water.

(*iii*) *Lipids*[19]

The calcareous shell of articulate brachiopods contains some lipid. In the organic web of rhynchonellides, *Notosaria* and *Hemithiris* (free of caeca), lipids are about 5% of total organics—neutral lipids 40%, chiefly cholesterol or its ester, and polar lipids 60%. The polar lipids, by silicic acid column fractionation with chloroform–methyl alcohol mixtures, have yielded glycolipids and phospholipids—a high proportion of kephalins chiefly phosphatidyl serine, small amounts of lecithins, and probably small amounts of sphingomyelin and phosphoinositide.

Terebratulide calcareous shell, containing caecal outgrowths of the mantle epithelium (see p. 765) has thus an additional source of lipid. Organic web lipids and those of mantle epithelium cell walls have been compared. The latter (on the order of 20% of the mantle weight, whole cells) yielding 30% neutral lipids (cholesterol and its ester) and 70% polar lipids—glycolipids and phospholipids with composition very similar to that of organic web phospholipids. By comparison, intestinal microvillus plasma membrane lipids (rat)[63] have 20% neutral lipids (70% cholesterol), 30% phospholipids (chiefly phosphatidylethanolamine) and 50% other polar lipids, possibly glycolipids.

CONSTITUENTS OF BRACHIOPOD SHELLS XIII

In the organic web of *Notosaria* about half (in alternating layers) is of triple-unit structure. Most (if not all) of the lipid is probably located in the triple-unit sheet (with its membrane-like appearance) yielding at least 10% lipids for this structure—low compared with erythrocyte membranes $(20\text{--}40\%)$[26,28] and about 38% for intestinal microvillus membrane (rat)[63]. This low value suggests that the brachiopod shell 140-Å triple-unit sheet has molecular organisation different from that of a bilipid leaflet.

It would be difficult to account for this 140-Å triple-unit structure in terms solely of spacing of a lipid bilayer separating protein layers. For although the chain lengths of the lipid fatty acids vary considerably, the molecules pack into a uniform space in a lipid bilayer giving a constant phosphorus to phosphorus distance[21] of 51.5 Å. A configuration has been proposed[23] in which even the large sphingosine–lipid molecules can be accommodated within this distance: saturated chains being straight but *cis*-unsaturated chains coil. Sphingomyelin and cerebrosides, at their fullest extent uncoiled, would be long enough to give this brachiopod spacing[64], but these compounds are not present in sufficient proportion. Hydration may be important in this context, for X-ray data[64] suggest that lipid molecules, in particular kephalins, have a longer effective length when wet.

4. Shell formation

(a) Growth processes

The shell is laid down by a group of mantle cells whose secretion pattern progressively changes with the growth of the shell[5–8]. The mantle is a sheet of active tissue, a single layer of ectodermal epithelial cells (lining the anterior part of each valve), turned back at the shell margin to give an outer epithelium in contact with the inner surface of the shell and an inner epithelium lining the mantle cavity which is in direct communication with the sea. Between the outer and inner epithelium is a thin layer of connective tissue. The mantle edge, the junction of outer and inner epithelium, is folded in (in most brachiopods) to give an inner and an outer lobe. The groove between the two lobes is the generative zone, a centre of active growth. The animal grows by cell division in this zone (Fig. 1). These cells are, at first, part of the inner surface of the outer mantle lobe, then as more cells are formed, are pushed forward to become the tip of the outer mantle lobe, and finally with further growth, become part of the general surface of the outer epithelium.

(i) The periostracum[8]

The periostracum is formed by cells of the inner surface of the outer mantle lobe (Fig. 1). First a mucus layer is secreted by a newly formed cell to reach 100-nm thickness by the time further new cells have carried this cell to the third or fourth position in the generative zone. This cell now begins to secrete the periostracum beneath the mucus layer. The outer bounding membrane is first laid down, then is pushed away from the cell surface by secretion of protein–polysaccharide material (the main bulk of the periostracum) which accumulates up to about 600 nm, then is sealed off by the inner bounding membrane (Fig. 2). Formation of the periostracum appears to be rapid (as seen in young *Notosaria*) for it is completely formed before the cell reaches the tip of the outer mantle lobe.

Anomalies in the structural pattern may be caused by repetition of certain secretion phases or by sudden forward movements or retreats of the mantle tip, resulting in multiplication of segments of the outer membrane or in breaks in the continuity of either membrane.

The mucus layer which precedes formation of the brachiopod periostracum is analogous to the *pellicle*, a 40-nm electron-dense layer (protein), preceding periostracum formation in the lamellibranch *Macrocallista maculata*[65]. The pellicle acts as a barrier between cells of the middle mantle fold and secretion (of periostracum) from cells of the outer fold—a function which holds also for brachiopods. Extraneous coats of a similar nature appear to be present in greater or lesser extent on almost all cells and may take part in such important mechanisms as pinocytosis[66]. Protein and both acid and neutral mucopolysaccharides have been detected in them.

(ii) The calcareous primary layer[5,6]

The calcareous primary layer is formed by cells at the tip of the outer mantle lobe (Fig. 1). When a cell reaches the tip of the outer mantle lobe its secretion pattern changes to deposit calcium as fine fibrous calcite on the under surface of the periostracum. This may act as a seeding surface for future crystallisation. The calcite crystals are aligned with their c-axes normal to the shell surface. This layer is secreted by a small number of cells at the mantle tip (about 15 in *Notosaria*) and its thickness is regulated by the number of rows of these cells concerned with its secretion. It is deposited without regard to cell boundaries, bearing no imprint of cell outlines, but has a strongly pitted appearance, perhaps caused by intrusion of microvilli of the secreting cells or by small exudations of protein. The calcite of this layer represents unspecialised crystallisation with calcium-rich fluid contained between mantle and periostracum.

(iii) The calcareous secondary layer[5,6]

The calcareous secondary layer is formed by cells of the general surface of the outer mantle epithelium (Fig. 1). When cells have been pushed forward from the mantle tip to become part of the general outer surface of the mantle, the pattern of secretion changes again. Organic material (appearing in electron micrographs as a triple-unit layer) is secreted from the anterior boundary of a cell, held temporarily by desmosomes to an arcuate zone of the membrane. Beneath this sheet calcite is deposited beside the posterior zone of the plasma membrane, initially in crystallographic continuity with that of the primary layer. The calcite fibre is laddered by fine organic layers possibly representing oscillations in calcite deposition.

Growth of the fibres starts in a direction normal to the shell edge for a short distance, then turns to run parallel to the valve margin, each fibre tending to grow in a spiral arc directed clockwise in the right half of the valve and counterclockwise in the left. The fibres are inclined at 10° to the primary layer due to accumulation of calcite solely on the posterior surface of the cell. The growth of each fibre is controlled by the concave outer surface of a cell causing the proximal face of the fibre to be convexly lobate with an outline determined by the boundary of the triple-unit layer. The inner surface of the secondary shell, by contrast with the primary, thus shows the characteristic pattern of the shell mosaic—a protein–calcite reflection of the cell boundaries of the outer surface of the epithelium (Fig. 4B). Like the calcite of the primary layer these calcite fibres have their *c*-axes aligned normal to the shell surface but oblique to the length of the fibre.

The three regions of the shell are thus laid down by the same cells as they grow progressively away from the generative zone to form different parts of the mantle, each region of the shell being deposited in turn as the cells arrive at a particular location.

(b) Calcification

The mechanisms of calcification, whether carbonate or phosphate, in relation to organic tissue, still present one of the outstanding problems in biology. Aspects of the subject have been recently reviewed by Glimcher and Krane[67], by Eastoe[36] and by Wilbur and Simkiss[41]. Recent work on brachiopods could be instructive, for the phylum includes both carbonate and phosphatic calcification, and their shell proteins are becoming known, hydroxyproline-containing proteins being associated with phosphatic calcification[2,12].

Calcification in brachiopods takes place extracellularly in the space between the mantle epithelium and the periostracum. With articulates the precise correspondence between the calcite fibres of the inner calcite (bounded by protein material) and the epithelial cell, through the shell mosaic, suggests that at least here the mantle cells may be responsible for providing the Ca^{2+} or the anions or both. The orientation of the c-axis throughout the curving length of these fibres is normal to the shell surface, but the arcuate rungs of protein material seen along these fibres suggest pauses in deposition perhaps reflecting periodic ionic depletion followed by recovery to supersaturation, as is frequent in crystal deposition.

The basic chemical processes of calcareous deposition are simple, but it is not clear how they are regulated by the organic system to yield replicable heritable mineralized shapes. The very universality of these two types of calcification through so wide a range of biochemical systems and micro-environments suggests that the processes of biological calcification are basically simple, depending as much upon the properties of the calcium atom and its ionic forms (and particularly its states of hydration) in relation to the negative ions, as upon specific biochemical circumstances (though Eastoe sees it as a complex chain of events). A mechanism for concentrating the effective ionic strength of Ca^{2+} (either membranous or creating an ionic gradient) and perhaps controlling negative ions must be universally needed.

There are persistent suggestions that the structural proteins of the shell have some determining influence upon the siting of initiation of carbonate and phosphate deposition, beyond simple membranous control of Ca^{2+} and negative ion concentrations. Such suggestions have yet to be substantiated in detail, but the information for controlling the inorganic shaping has to be communicated by some means from the organic molecules to the inorganic growth in which heritable shapes are also manifest; this may need more than a mere series of constraints to the directions of growth. Isolated calcite rhombs on patches of protein on the under surface of the periostracum have been observed in electron micrographs[6] of the brachiopods *Notosaria* and *Waltonia* and similar crystals in analogous positions on other organisms[67,68] which some workers have considered the first seeds of calcite of the calcareous primary layer.

It may be significant that among the brachiopods, as widely through nature, those structural proteins containing hydroxyproline and hydroxylysine run with phosphatic calcification, though it is not yet clear whether this is a direct connection or merely a parallel effect from a deeper determinant, such as the operation of proline- and lysine-hydroxylating processes, which might interfere with enzyme molecules such as carbonic anhydrase and so disturb the balance of other conditions such as local CO_2 concentrations. The observation that hydroxyproline and hydroxylysine are virtually absent from immature dental enamel[36] in which much hydroxyapatite has already been deposited might suggest that the relation is not direct.

Epitaxial growth of calcite or phosphate on the structural protein surface is sometimes invoked as a determinant, without due consideration of the molecular, atomic and ionic spatial and electron-density interrelations involved. Configurations on a protein molecule might possibly favour the dehydration of Ca^{2+} requisite for crystallisation[69], but holding of a Ca ion in an ionic linkage (let alone chelation) is unlikely to lead directly to crystal formation because of the deformation of the electron-density patterns.

The calcium phosphate system, though with molecular, ionic, solubility and crystal-lattice properties markedly different from those of the calcium carbonate system, nevertheless provides an alternative source of exoskeletal and skeletal material strength.

Calcium phosphate is found in inarticulate brachiopod shell (*Lingula*) as hydroxyapatite crystallites (needles) up to about 1000 Å long (comparable with those of bone), which appear to have developed from round particles about 50 Å across in an electron-dense organic background which disappears as the crystallites increase in size[70]. It is possible that initial deposition is as non-crystalline calcium phosphate (perhaps organically associated), which acts as an easily mobilized reservoir for Ca^{2+} and phosphate providing a steady-state condition of supersaturation during conversion to crystalline hydroxyapatite[71].

Phosphatic calcification, in general, is associated widely in nature with hydroxyproline-containing structural proteins, the collagens of bone or the less highly ordered molecules of inarticulate brachiopod shell. These proteins may favour crystal nucleation or conversion to hydroxyapatite, though this has yet to be clearly shown, and the subject requires detailed physico-chemical investigation at the molecular level. Biological phosphatic calcification at Ca^{2+} and phosphate ionic concentrations less than half those needed *in vitro*[72], while without much doubt in part biochemically regulated, may still turn out to be conditioned also by localized differential phosphate and carbonate ionic concentrations.

5. Concluding remarks

This survey of Brachiopod exoskeletal biochemistry has raised several matters of wider biological significance. It has shown that a taxonomy with apparently phylogenetic implications can reside in the amino acid composition of structural proteins of shell, but that functional factors may take precedence in mantle and muscle[12]. It has shown also that with the phosphatic calcification of Inarticulates the shell protein contains hydroxyproline and hydroxylysine, a relation widely observed in nature. These proteins are not otherwise like collagens (as molecules they are less highly ordered); collagens are however found in these organisms, for example in connective tissue and pedicle. It is still not clear how far these structural proteins are determinants in calcification generally or in transmitting heritable shaping from the organic to the inorganic shell structure; studies on brachiopods with their uniquely close relations, within the phylum, between carbonate and phosphate calcification, may help to elucidate the general mechanisms of calcification.

Electron microscopy has displayed in detail the microstructure of the exoskeleton in relation to the associated soft parts, and this has now to be investigated at the molecular level, to define for instance the true nature of the triple-layered membranous appearance of the shell organic web.

The molecular organisation of these shell structural proteins now needs to be investigated comprehensively by physical methods—wide-angle X-ray diffraction, infrared absorption, and optical rotation studies—and chemically by amino acid sequencing and showing protein molecular relations with carbohydrates and lipids, and also with the chitin found in inarticulate phosphatic shell. One difficulty in understanding the true nature and role of these intractable structural proteins is that procedures for taking them into polar solutions suitable for studying their properties tend to disrupt their native molecular configuration as they exist in the shell structure. Further precise differentiation of the molecular species constituting these protein webs is still needed.

Investigations are needed into the finer details of such structural protein biosynthesis in relation to ribosome clusters, and the extent to which the macromolecules may be finally assembled from subunits (as perhaps suggested by some recent sedimentation data[73]), and also on the mechanisms whereby such intracellularly produced protein material is transferred to be incorporated in the shell structure.

References p. 783

Recent work[19] indicates that two brachiopod groups (Terebratulida and Craniacea) have the same N-terminal amino acid (phenylalanine) in the polypeptide chains of their shell proteins (taken into solution under comparatively mild conditions: in 0.1 M $NaHCO_3$–7M urea at room temperature). This might prove to be a phylogenetic characteristic of brachiopod shell proteins, with perhaps methionine as the initiating amino acid in protein synthesis removed at an early stage (*cf.* the analogous position of N-terminal valine in the α, β, and δ chains of hemoglobin[74]).

REFERENCES

1   M. JOPE, in R. C. MOORE (Ed.), *Treatise on Invertebrate Palaeontology*, Vol. H, *Brachiopoda*, University of Kansas, Kansas, 1965, pp. 156–164.
2   M. JOPE, *Comp. Biochem. Physiol.*, 20 (1967) 593.
3   A. WILLIAMS AND A. J. ROWELL, in R. C. MOORE (Ed.), *Treatise on Invertebrate Palaeontology*, Vol. H, *Brachiopoda*, University of Kansas, Kansas, 1965, pp. 57–155.
4   A. WILLIAMS, *Biol. Rev. Cambridge Phil. Soc.*, 31 (1956) 243.
5   A. WILLIAMS, *Nature*, 211 (1966) 1146.
6   A. WILLIAMS, *Evolution of the Shell Structure of Articulate Brachiopods*, Palaeontological Soc., London, 1968, pp. 1–30.
7   A. WILLIAMS, *Lethaia*, 1 (1968) 268.
8   A. WILLIAMS, *Nature*, 218 (1968) 551.
9   G. OWEN AND A. WILLIAMS, *Proc. Roy. Soc. (London)*, Ser. B, 172 (1969) 187.
10 A. WILLIAMS, unpublished work.
11 A. WILLIAMS AND A. D. WRIGHT, *Shell Structure of the Craniacea and Other Calcareous Inarticulate Brachiopoda*, Palaeontological Soc., London, 1970, pp. 1–51.
12 M. JOPE, *Comp. Biochem. Physiol.*, 30 (1969) 209.
13 M. JOPE, unpublished work.
14 H. FERNÁNDEZ-MORÁN, *Circulation*, 26 (1962) 1039.
15 H. FERNÁNDEZ-MORÁN, T. ODA, P. V. BLAIR AND D. E. GREEN, *J. Cell Biol.*, 22 (1964) 63.
16 T. ODA AND S. SEKI, *J. Electron Microscopy*, 14 (1965) 210.
17 T. ODA, *J. Electron Microscopy*, 17 (1968) 174.
18 F. S. SJÖSTRAND, in J. JÄRNEFELT (Ed.), *Regulatory Functions of Biological Membranes (BBA Library, Vol. 11)*, Elsevier, Amsterdam, 1968, pp. 1–20.
19 M. JOPE, to be published.
20 A. WILLIAMS, *Smithsonian Inst. Misc. Collections*, 1970.
21 J. B. FINEAN, *Circulation*, 26 (1962) 1151.
22 W. STOECKENIUS, *J. Cell Biol.*, 12 (1962) 221.
23 F. A. VANDENHEUVEL, *Ann. N.Y. Acad. Sci.*, 122 (1965) 57.
24 D. E. GREEN AND J. F. PERDUE, *Proc. Natl. Acad. Sci. (U.S.)*, 55 (1966) 1295.
25 D. E. GREEN AND J. F. PERDUE, *Ann. N.Y. Acad. Sci.*, 137 (1966) 667.
26 A. H. MADDY, *Intern. Rev. Cytol.*, 20 (1966) 10.
26a M. JOST AND D. D. JONES, *Can. J. Microbiol.*, 16 (1970) 159.
27 K. E. WOLFARTH-BOTTERMANN, *Protoplasma*, 53 (1961) 259.
28 B. ROELOFSEN, J. DE GIER AND L. L. M. VAN DEENEN, *J. Cell. Comp. Physiol.*, 63 (1964) 233.
29 A. P. VINOGRADOV, *Sears Foundation for Marine Research, Mem.*, 2 (1953) 647.
30 S. KAWAGUTI, *Taihoku Imp. Univ., Fac. Sci. Agr., Mem.*, 23 (1941) 95.
31 Spectrographic analyses done by E. DONALDSON, Queen's University, Belfast, 1969.
32 L. SCHMELCK, in H. FRIELE AND J. A. GRIEG (Eds.), *Norske Nordhavs Expedition (1876–1878)*, Vol. 7, *Zoology, Mollusca III*, 1901, p. 129.
33 G. V. POTAPENKO, *Trudy Nauch.-issled. Inst. Fiz. Kristall.*, 3 (1925) 16.
34 V. M. GOLDSCHMIDT AND C. PETERS, *Nachricht. Gesell. Wiss. Göttingen, Math.-Phys. Kl.*, (1932) 528.
35 W. D. ARMSTRONG, personal communication, 1962, in D. McCONNELL, *Geol. Soc. Am., Bull.*, 74 (1963) 363.
36 J. E. EASTOE, *Calc. Tiss. Res.*, 2 (1968) 1.
37 M. JOPE, *Comp. Biochem. Physiol.*, 30 (1969) 228.
38 G. F. ELLIOTT, A. F. HUXLEY AND T. WEIS-FOGH, *J. Mol. Biol.*, 13 (1965) 791.
39 D. C. PHILLIPS, *Sci. American*, 215 (1966) 78.
40 S. KOBAYASHI, *Biol. Bull.*, 126 (1964) 414.

41 K. M. WILBUR AND K. SIMKISS, in M. FLORKIN AND E. H. STOTZ (Eds.), *Comprehensive Biochemistry*, Vol. 26A, Elsevier, Amsterdam, 1968, pp. 229–295.
42 V. R. MEENAKSHI, P. E. HARE AND K. M. WILBUR, personal communication in ref. 41, K. M. WILBUR AND K. SIMKISS, p. 266.
43 E. T. DEGENS, D. W. SPENSER AND R. H. PARKER, *Comp. Biochem. Physiol.*, 20 (1967) 553.
44 E. T. DEGENS AND D. W. SPENSER, *Woods Hole Oceanogr. Inst.*, Ref. 66-27 (1966) unpublished manuscript.
45 S. O. ANDERSEN, in G. R. TRISTAM (Ed.), *The Structure and Function of Connective and Skeletal Tissue*, Butterworth, London, 1965, pp. 105–109.
46 R. DENNELL, *Biol. Rev.*, 33 (1958) 178.
47 S. UDENFRIEND, *Science*, 152 (1966) 1335; R. L. MILLER AND S. UDENFRIEND, *Arch. Biochem. Biophys.*, 139 (1970) 104.
48 B. GOLDBERG AND H. GREEN, *Nature*, 221 (1969) 267.
49 L. GLASER AND D. H. BROWN, *J. Biol. Chem.*, 228 (1957) 729.
50 A. TELSER, H. C. ROBINSON AND A. DORFMAN, *Proc. Natl. Acad. Sci. (U.S.)*, 54 (1965) 912.
51 K. M. RUDALL, in J. W. L. BEAMENT, J. E. TREHERNE AND V. B. WIGGLESWORTH (Eds.), *Advan. Insect Physiol.*, 1 (1963) 257–312.
52 K. M. RUDALL, in G. R. TRISTRAM (Ed.), *The Structure and Function of Connective and Skeletal Tissue*, Butterworth, London, 1965, pp. 191–196.
53 K. M. RUDALL, in M. BROWN AND J. F. DANIELLI (Eds.), Fibrous Proteins and their Biological Significance, *Symp. Soc. Exptl. Biol.*, 9 (1955) 58.
54 N. E. DWELTZ, *Biochim. Biophys. Acta*, 51 (1961) 283.
55 C. JEUNIAUX, *Chitine et Chitinolyse, un Chapitre de la Biologie Moléculaire*, Masson, Paris, 1963, pp. 54 ff.
56 C. JEUNIAUX, *Arch. Intern. Physiol. Biochim.*, 72 (2) (1964) 329.
57 H. FÜLLER, *Zool. Anz.*, 177 (1966) 296.
58 A. G. RICHARDS, *The Integument of Arthropods*, University of Minnesota, Minneapolis, 1951, pp. 174–177.
59 R. DENNELL, *Proc. Roy. Soc. (London)*, B, 133 (1946) 348.
60 R. H. HACKMAN, *Australian J. Biol. Sci.*, 13 (1960) 568.
61 M. M. ATTWOOD AND H. ZOLA, *Comp. Biochem. Physiol.*, 20 (1967) 993.
62 K. M. RUDALL, *Biochem. Soc. Symp.*, 25 (1965) 83.
63 G. G. FORSTNER, K. TANAKA AND K. J. ISSELBACHER, *Biochem. J.*, 109 (1968) 51.
64 R. S. BEAR, K. J. PALMER AND F. O. SCHMITT, *J. Cell. Comp. Physiol.*, 17 (1941) 355.
65 G. BEVELANDER AND H. NAKAHARA, *Calc. Tiss. Res.*, 1 (1967) 55.
66 S. ITO, *J. Cell Biol.*, 27 (1965) 475.
67 M. J. GLIMCHER AND S. M. KRANE, in B. S. GOULD (Ed.), *Treatise on Collagen*, Vol. 2B, Academic Press, New York, 1968, pp. 137–251.
68 K. M. WILBUR AND N. WATABE, *Proc. Intern. Conf. Trop. Oceanogr.*, Univ. of Miami Institute of Marine Sciences, Miami, Florida, 1967, pp. 147–148.
69 B. E. DOUGLAS AND D. H. MCDANIEL, *Concepts and Models of Inorganic Chemistry*, Blaisdell, New York, 1965, p. 125.
70 P. G. KELLY, P. T. P. OLIVER AND P. G. E. PAUTARD, *Proc. Second European Symp. on Calcified Tissue*, University of Liège, 1965, p. 337.
71 E. D. DANES, I. H. GILLESEN AND A. S. POSNER, in H. S. PEISER (Ed.), *Crystal Growth:* suppl. to *J. Physics and Chemistry of Solids*, Pergamon, Oxford, 1967, p. 373.
72 B. N. BACHRA, *Ann. N.Y. Acad. Sci.*, 109 (1963) 251.
73 A. J. BAILEY, in M. FLORKIN AND E. H. STOTZ (Eds.), *Comprehensive Biochemistry*, Vol. 26B, Elsevier, Amsterdam, 1968, p. 404.
74 R. JACKSON AND T. HUNTER, *Nature*, 227 (1970) 672.

Chapter XIV

Dental Enamel

JOHN E. EASTOE

Department of Dental Science, Royal College of Surgeons of England, London (Great Britain)

1. Introduction

Dental *enamel* is here considered as the hard, coherent, highly mineralized material which covers the incisal and occlusal surfaces of the teeth in higher vertebrates. Its unique structure gives enamel mechanical properties which enable it to resist abrasive wear and the shocks of impact involved in the mastication of food. The chemical nature of enamel frequently permits it to survive in substantially unchanged form after death, longer than any other material in the body, thereby providing palaeontological information about its possessor.

The layer of enamel is firmly attached to *dentine* which lies deeper within the tooth (Fig. 1). This tissue is also mineralized but differs from enamel in containing cytoplasmic processes and having an organic matrix which consists largely of collagen. Dentine which, in turn, encloses an unmineralized zone known as the pulp in the centre of the tooth, resembles bone in composition although it is distinct histologically. The outer surfaces of both enamel and dentine may be wholly or partially covered by another mineralized tissue, *cementum*, although enamel is more frequently exposed as it is in normal human teeth. Cementum has a collagenous matrix and resembles bone rather closely in both composition and histological structure.

Enamel is laid down by a single layer of cells, the *ameloblasts*, which are considered to be of ectodermal origin. The enamel has been shown to lie exterior to the plasma membrane[1,2] of the ameloblasts which are found on its outer surface, remote from the dentine (Fig. 1). Thus enamel resembles

[785]

a secretion rather than a keratinised tissue, in which mechanical components of the epithelium remain within the cells which produce them. Dentine is produced by a layer of modified fibroblasts of mesodermal origin known as odontoblasts. At an early stage of tooth formation, the ameloblasts and

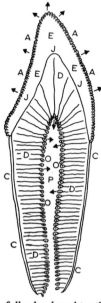

Fig. 1. Longitudinal section of a fully developed tooth showing the relative positions of enamel (E), dentine (D), the dentine–enamel junction (J), cementum (C) and pulp (P). At this stage, the ameloblasts (A) would have degenerated and worn away, but they are shown in their final positions on the surface of the fully developed enamel. The arrows indicate the directions in which the ameloblasts and odontoblasts (O) moved when laying down enamel and dentine during tooth formation.

odontoblasts are in close proximity at an interior surface which becomes the dentine–enamel junction, a fixed reference datum for all subsequent stages of development. The odontoblasts and ameloblasts move respectively inwards and outwards away from the dentine–enamel junction in opposite directions, remaining as single layers of cells which lay down dentine and enamel behind them.

In the fully formed tooth, the odontoblasts remain on the inner surface of the dentine, lining the walls of the pulp cavity. They are alive and partici-pate to a limited extent in repair of dentine, should it subsequently become damaged by disease or accident. The ameloblasts, on the other hand, are on the outside of the tooth and degenerate or otherwise change their characteristics in the later stages of tooth development. In any event, after eruption, their remains are rapidly worn away and replaced by an *acquired pellicle* derived from salivary and microbiological sources. The enamel of erupted teeth is thus subject only to physico-chemical processes, beyond the direct cellular control of the animal.

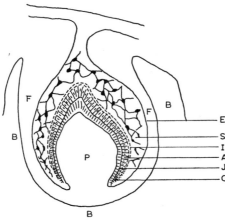

Fig. 2. Longitudinal section of a tooth bud immediately prior to deposition of enamel and dentine showing pre-ameloblasts (A), dentine–enamel junction (J), preodontoblasts (O), stratum intermedium (I), stellate reticulum (S), outer dental epithelium (E), pulp (P), tooth follicle (F) and the bone of the jaw (B).

During early stages of development, the enamel, within the tooth bud (Fig. 2), is surrounded by a system of tissues known as the enamel organ. It consists of layers of epithelial cells known as the *inner dental epithelium* (*pre-ameloblasts*) and the *stratum intermedium*, a broader zone of more widely spaced cells, the *stellate reticulum* and, on the exterior surface of the enamel organ, the *outer dental epithelium*. The functions of the enamel organ are probably concerned with control of the development and nutrition of the growing tooth tissues, which the organ encloses. The mechanism is not yet altogether clear but histochemical investigations have shown the system to be an active producer of enzymes. The intercellular spaces of the stellate reticulum are rich in polysaccharide, which by means of changes in its water content can exert changes in pressure on the neighbouring tissues, perhaps thereby controlling the exact shape of the forming dentine–enamel junction, which subsequently determines the shape of the particular tooth.

True dental enamel, defined as a product of ectodermal ameloblasts, appears to be confined to reptiles and mammals. The teeth of elasmobranchs, fishes and amphibia have a glassy outer covering known as *enameloid*, which differs from true enamel in being formed from mesodermal cells and in containing collagen when newly formed[3]. Thus enameloid, which covers not only the teeth but also the scales of lower vertebrates, has affinities with dentine but differs from it in so far as its collagenous matrix gradually undergoes "reduction" and is replaced almost completely by calcium salts[4].

The evolutionary origins of enamel are still far from clear. No closely related system has yet been demonstrated in invertebrates, although the shell of *Lingula unguis*[5] and, to a lesser extent, the mineralized byssus of certain molluscs share some characteristics with enamel. Poole states that the teeth of all reptiles, including the primitive cotylosaurs, are covered by true enamel and considers that it may have arisen in the lung-fish ancestors of amphibians. Levine *et al.*[6] separated an insoluble protein from the mature enameloid of the hammer-head shark which was similar in composition to the insoluble protein from mature mammalian enamel. This finding could, if correct, indicate additional similarities between true enamel and enameloid and perhaps relate the origin of mature enamel protein phylogenetically to systems involving loss of collagen[3].

Several textbooks[7-9] and reviews[10-14] are available which discuss specific aspects of the biology, structure and chemistry of enamel in greater detail than space permits here.

2. Characteristics and constituents of enamel

(a) Some features of enamel structure

An essential characteristic of mammalian enamel is that it shows structural organisation at two orders of size. (*1*) Basically, at the order of size represented by the electron microscope, enamel is a two-phase system consisting of discrete crystals of hydroxyapatite in a continuous phase consisting of organic material associated with water. (*2*) At a histological level, observable with the optical microscope, mammalian enamel is organised in an elaborate but clearly recognisable pattern the unit of which is the enamel rod or prism. These two levels of organisation respectively reflect the contributions of molecular processes and the overall activity of cells to building up the entire structure. Reptile enamel, which is comparatively thin, has a similar structure to that of mammals at the lower order of size but its histological elaboration is less developed, structure being more continuous than in mammals. Mammalian teeth are present in the mouth for long periods and have much thicker enamel which, because of the hard brittle nature of hydroxyapatite, would shatter on impact if it had a continuous histological structure. The prismatic pattern thus appears to provide increased elasticity.

(i) Enamel as a two-phase system

Electron micrographs of mature enamel show that most of the volume is occupied by inorganic crystallites[15,16]. These are discontinuous but are in such close contact that the second phase consisting of water and organic matter occupies only narrow gaps between the crystals. At early stages of development the crystals form long thin ribbons or plates[17] and a considerably higher proportion is occupied by the organic phase. The two-phase concept is important in relation both to the growth of discrete apatite crystallites in an organic continuum during enamel maturation and to the concept of "spaces" in mature enamel. These spaces are responsible for the microporous properties of mature enamel and have been demonstrated optically with polarized light after imbibition with various liquids having molecules of different sizes, the spaces becoming progressively enlarged in dental caries[18].

(ii) Histological structure

Enamel has been widely studied with the optical and polarizing microscope for many years and a "classical" concept of its structure was built up in which the features observed were attributed directly to specific histological elements. This caused some confusion because the highly refractive hydroxyapatite gives rise to a variety of optical effects which cannot be directly interpreted structurally, without further information. Consequently some of the ideas of enamel structure based only on observations of ground sections have been drastically revised as a result of recent research. The following account deals only with those morphological features which have some bearing on chemical structure.

The enamel prism or rod survives as the most obvious histological entity. Its width in two dimensions is approximately the same as an ameloblast, but it is extremely long, possibly extending from the dentine–enamel junction to the enamel surface. Neighbouring prisms are parallel but may show shifts of direction in some species. It was originally considered that each ameloblast was responsible for the formation of one prism only, but Boyde[19] has shown how each cell may contribute to more than one prism, while each prism may receive material from more than one ameloblast, the arrangement varying somewhat between species.

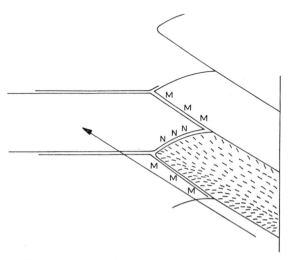

Fig. 3. The mineralizing surface of prismatic enamel showing prisms and crystallites on the right and ameloblasts to the left. The arrow indicates both the prism direction and the direction in which the ameloblasts move relative to the stationary enamel. Each amelo-blast contributes to two prisms (surfaces MMM and NNN). Crystallites at NNN are perpendicular to the mineralizing surface, while those at MMM are somewhat displaced from a direction perpendicular to the surface by the sliding movement between the ameloblasts and mineralizing front[19,20].

The hydroxyapatite crystals are mostly aligned with their long axes parallel to one another and at a small angle to the prism direction (see p. 803) but near the edge of each prism they deviate progressively, those ends of the crystallites which point towards the enamel surface being turned towards the prism boundary. Boyde[19] relates the orientation of crystallites to the shape of the mineralizing front which is typically like a "picket" fence (Fig. 3), the ameloblasts fitting into the gaps. If each crystallite were formed in a direction perpendicular to the surface at its point of formation, the orientation within the prism would be accounted for.

The boundary between prisms which is distinct in the microscope was at one time attributed to an organic "prism sheath" composed of interprismatic substance which cemented prisms together. As a result of recent investigations with the polarizing and electron microscopes, it is now clear that discrete structures such as the prism core, prism sheath and interprismatic substance do not exist as separate entities[20], and that the prism boundaries are rendered visible almost entirely by the abrupt change in crystallite orientation which occurs there[21]. Nevertheless, electron micrographs of the earlier stages of enamel maturation show that the region of the boundary of adjacent prisms contains fewer crystallites and consequently more of the continuous organic phase than within the prisms[22]. It is therefore probable that when maturation is complete, a higher concentration of residual diffuse organic material will be found near prism boundaries than inside the prisms. This is a consequence of the two-phase system not being uniformly or randomly distributed but laid down according to the larger-scale histological pattern. As a result, while prism sheaths are not definite organelles, the highest local concentrations of organic matter and hence the largest pores are likely to follow the neighbourhood of prism boundaries, though the microporous system ramifies between all crystallites of the two-phase system. The size of pores in enamel covers a very broad range from 3 to 5000 Å diameter[23].

(b) Overall changes during mineralization and maturation

Newly secreted enamel is quite soft and it remains friable for some time after it is laid down. There is a continuous increase in the degree of mineralization resulting in the attainment of a final extreme hardness before the tooth erupts. Several processes and stages take place successively during enamel development. Enamel formation involves *secretion* of the organic phase by ameloblasts followed by its *mineralization*; this in turn concerns two processes, *crystallite formation*, which occurs soon after the organic phase is produced, and *crystal growth*, which takes place in several stages (see p. 803). The entire process by which the newly secreted organic phase becomes very highly mineralized is referred to as *maturation*. Knowledge of how this occurs has greatly contributed to an understanding of the structure of mature enamel at all orders of size.

The first quantitative study of enamel maturation was published in 1942 by Deakins[24] who measured the contents of inorganic matter, water and

organic matter in dissected pieces of pig enamel at various stages of mineralization. These changes were referred to the increasing density of the
enamel which enabled the amount of each constituent to be calculated per
unit volume of enamel. Water was determined as weight lost on heating
enamel to 110°, organic matter as weight lost between 110° and 700° and
inorganic matter as ash at 700°. A correction was made to the value for
organic matter on account of loss of carbon dioxide on heating carbonates.

As maturation proceeded the enamel increased in density and became
harder. Between densities of 1.45 and 1.75 the enamel was sufficiently soft
to be cut without it fracturing, from 1.75 to 2.25 it could be cut but was
friable, while from 2.25 to the highest observed density of 2.76 it was too
hard to cut with a scalpel. The composition of the enamel samples at the
extremes of the density range, representing the beginning and end of the
maturation process, are summarised in Table I. The minimum value for
organic matter in mature enamel corresponds to 1.8 weight per cent. This
is substantially higher than more recent values for total organic matter and
it would appear that the figure includes some firmly bound water, retained
by enamel at 110° but lost below 700°. It seems likely that if the error from
this source were corrected the fall in inorganic content would probably
continue above a density of 1.65 but at a decreased rate.

TABLE I

COMPOSITION OF PIG ENAMEL AT THE LOWEST AND HIGHEST DEGREES OF
CALCIFICATION[24]

Density	Lowest calcification very soft 1.45			Highest calcification very hard 2.76			Change in volume
	mg/mm^3	% by wt.	vol.[a]	mg/mm^3	% by wt.	vol.[a]	
Inorganic	0.54	37.0	0.16	2.62	95.0	0.82	+0.66
Organic	0.27	19.0	0.20	0.05	1.8	0.04	−0.16
Water	0.64	44.0	0.64	0.12	4.3	0.12	−0.52
Totals	1.45	100.0	1.00	2.79	101.1	0.98	

[a] Volume in mm³ occupied by each constituent, calculated assuming densities of 3.18
for inorganic and 1.31 for organic material.

It is clear that there is a continuous gain of inorganic material during
maturation and a simultaneous and reciprocal loss of organic matter and
water. Furthermore these losses are not simply the result of dilution by

TABLE II

NITROGEN AND AMINO ACID CONTENT AT DIFFERENT STAGES OF ENAMEL FORMATION[a,28]

Stage of enamel formation	1 Early matrix	2 Complete matrix	3 Early mineralization	4 Late mineralization	5 Tooth erupted
Mean density of wet enamel	1.780	1.720	2.060	2.720	2.820
Amino acid (μmoles/100 ml)					
Aspartic	8.25	7.21	3.44	1.55	1.46
Glutamic	36.00	30.74	12.09	1.28	1.25
Serine	11.56	9.54	5.21	1.49	1.34
Threonine	6.50	5.94	2.42	0.36	0.35
Glycine	14.07	11.45	7.25	3.98	3.58
Alanine	5.75	4.66	2.14	1.57	1.36
Valine	8.56	8.69	3.53	0.36	0.38
Leucine	24.09	20.35	8.56	0.61	0.46
Isoleucine	8.50	7.42	3.16	0.36	0.37
Half-cystine	—	—	—	0.25	0.19
Methionine	13.78	11.45	4.00	—	—
Phenylalanine	6.23	5.72	2.70	0.29	0.26
Tyrosine	11.25	10.39	5.56	0.16	0.11
Proline	67.01	53.00	23.81	2.63	2.34
Hydroxyproline	0.76	0.64	0.74	1.08	1.04
Hydroxylysine	0.50	1.06	0.18	0.13	0.13
Lysine	3.75	3.39	1.95	0.52	0.48
Histidine	17.00	13.79	5.67	0.32	0.27
Arginine	4.51	4.03	2.32	0.88	0.62
Amide-N (mmoles/100 ml)	0.0400	0.0291	0.0145	0.0090	0.0001
Total amino acid					
mmoles/100 ml	0.248	0.210	0.095	0.018	0.016
g/100 g[b]	15.900	13.900	5.110	0.730	0.630
Total nitrogen					
moles/100 ml	0.341	0.284	0.127	0.031	0.028
g/100 g	2.680	2.310	0.863	0.159	0.144

[a] Undemineralized enamel samples. All values expressed on wet weight basis.
[b] Not corrected for water lost in peptide bond formation.

References p. 831

newly acquired inorganic salts. More evidence to support this resulted from the work of Stack[25,26] who separated by flotation the enamel of complete deciduous dentitions obtained from human foetuses and infants. It was shown that the total *weight* of protein present in the enamel of the whole dentition rose with time to a maximum, reached one month after birth, and then decreased to a final weight only one tenth of that present at the maximum. Stack[25,26] also found that in the enamel of various developing tooth crowns, where enamel deposition was still incomplete, the water and organic contents were proportional to one another. According to this, loss of water

TABLE III

AMINO ACID RESIDUES PER 1000 AT DIFFERENT STAGES OF ENAMEL FORMATION[a,28]

Stage of enamel formation	1 Early matrix	2 Complete matrix	3 Early mineralization	4 Late mineralization	5 Tooth erupted
Aspartic	33	34	37	86	91
Glutamic	144	145	130	71	78
Serine	46	45	56	83	84
Threonine	26	28	26	20	22
Glycine	56	54	78	221	224
Alanine	23	22	23	87	85
Valine	34	41	38	20	24
Leucine	96	96	92	34	29
Isoleucine	34	35	34	20	23
Half-cystine	—	—	—	14	12
Methionine	55	54	43	—	—
Phenylalanine	26	27	29	16	16
Tyrosine	45	49	49	9	7
Proline	268	250	256	149	146
Hydroxyproline	3	3	8	60	65
Hydroxylysine	2	5	2	7	8
Lysine	15	16	21	29	30
Histidine	68	65	61	18	17
Arginine	18	19	25	49	39
Amide per 1000 amino acid residues	160	139	156	500	504

[a] Undemineralized enamel samples.

and protein from enamel would appear to be simultaneous whereas the findings of Deakins[24] indicated that organic matter is lost rather earlier than water.

It is implied in the work of Stack[25] that protein is a major constituent in the organic phase of immature enamel. This is supported by the contemporary finding of Eastoe[27] who characterised the bulk protein of human foetal enamel, and showed that it accounted for approximately 20% of the weight of enamel. This value for protein was almost equal to that found by Deakins[24] for total organic matter of the youngest pig enamel examined. Partial fractionations of protein material based on differences in solubility suggested that more than one protein was present in young enamel. The first systematic separations of enamel proteins by chromatographic and electrophoretic methods were carried out by Burgess and Maclaren[28] who also carried out amino acid analyses of hydrolysates of whole enamel samples at different stages of mineralization (Table II). The results showed clearly that not only was there an overall loss of protein material from the enamel as mineralization proceeded, as shown by the fall in amino acid content from 15.9 to 0.63%, but also that different proteins were lost at different rates. This follows from the finding that the relative proportions of the various amino acids changed strikingly (Table III) as their contents in enamel all fell but at different rates. These changes are discussed in greater detail in section IV (p. 828).

The various pieces of evidence on the changes occurring in maturation are thus in good agreement that the overall process involves gain in inorganic matter and an absolute loss of water and organic matter by enamel. The main organic constituents of young enamel are proteins, which are differentially lost, at different rates from one another. Much of the chemistry of enamel is concerned with the nature of the substances involved in these changes and the mechanisms of the changes themselves.

(c) Chemical balance sheet for enamel

In order to characterise a biological system such as enamel, accurately, it is advantageous to be able to account for as much as possible of its weight in terms of constituents which have been identified chemically. Two difficulties arise with enamel, firstly immature enamel contains a mixture of several proteins, none of which has been completely characterised, secondly, in mature enamel, it is difficult to obtain a value for total water. This arises because water in enamel is held with various degrees of firmness from loosely bound water, lost below 100°, to water in closed pockets or possibly associated with the crystal lattice, which is retained at high temperatures.

(*i*) *Balance sheet for immature enamel*

Since the proportions of all main constituents alter with time there is no single fixed composition even in a particular species. The graphs of Deakins[24] are probably the best guide to the respective contents of the three main constituents, inorganic matter, organic matter and water at various stages, the values for the organic content being probably somewhat high and those for water slightly low (see p. 797). The organic fraction is probably accounted for almost entirely by protein, although this is an omnibus value for the several protein species present (see section IV, p. 805), the quantities of which have not yet been separately determined. Only very small quantities of other organic constituents have been detected, namely 0.23% carbohydrate[29] and 0.06% total phospholipid[30] (mainly lecithin and phosphatidylethanolamine with smaller amounts of sphingomyelin, phosphatidylinositol, phosphatidylserine, phosphatidic acid and cardiolipin).

(*ii*) *Balance sheet for mature enamel*

An attempt has been made in Table IV to account for all the weight of mature enamel in terms of identified constituents using reliable values from the literature. More detailed comparisons of the findings of different investigators and of local variations within human enamel are made by Brudevold and Söremark[31] for the inorganic components and Stack for organic ones[26]. Deakins' value for ash[24], after nitric acid treatment, together with the mean value quoted by Trautz for carbonate[32] correspond reasonably well with the overall value of 95% given by Brudevold and Söremark[31] for the inorganic phase.

Stack's value for total organic matter[38] by dichromate consumption during wet combustion may be slightly high, owing to the possibility of contamination of enamel, separated by flotation, with extraneous collagen. This may account for this value being substantially higher than the sum of the individual organic constituents in Table IV. A value of approximately 0.35% for true enamel protein is obtained by correction of total enamel protein for its collagen content[33]. The low molecular weight soluble proteinaceous material has been found both in outer and middle bovine[34] and human[33] enamel and is probably dispersed throughout mature enamel. The

TABLE IV

CHEMICAL BALANCE SHEET FOR MATURE ENAMEL

		per cent by wt.	
Total inorganic matter[31]			95.0
Ash, heated to 700° [24]		91.7	
Carbonate[32]		2.9	
Total organic matter by wet oxidation[38]			0.6
Total true enamel protein		0.35	
Low molecular weight material[33]	0.06		
Insoluble ribbons from inner molar enamel[33]	0.29		
Citrate[39]		0.02	
Lactate[39]		0.01	
Carbohydrate[36]		0.0016	
Fatty acid[37]		0.01	
Total water[31]			4.0
"Free" water, lost at 100°, *in vacuo*[40]		2.2	
"Bound" water, lost between 300 and 400° [41]		0.83	

insoluble ribbons[35] are found in the enamel near the dentine–enamel junction and may be restricted to human molar teeth. Citrate and lactate may be associated with the inorganic phase but are included with organic matter for comparative purposes. Values for citrate vary considerably[26] between investigators. The level of carbohydrate in mature enamel is very low, the main sugars present being galactose, glucose and mannose[36]. The total fatty acid content of enamel from human molars is 0.01% of which 19–25% is palmitate, 14–17% stearate and 20–21% oleate[37].

The total water content of mature enamel is uncertain because of difficulty in determining the most firmly bound water. Enamel reaches constant weight in one day at 100° *in vacuo* but requires one week at atmospheric pressure[40]. Little additional water is lost at 200° but there is a substantial loss[41] attributed to "bound" water between 300° and 400°. Probably some water is held even at this temperature and accounts for some of the loss in weight which occurs when the enamel is subsequently heated at 700°. A total value of around 4% for water in humidified enamel is therefore realistic but not easily verified. Casciani[42] found a very narrow proton line width (0.6 gauss) in enamel compared with a much wider one in mineral hydroxyapatite (2.5 gauss) which suggests loosely bound or "freely tumbling" water in enamel.

In Table IV only percentages by weight are given; in terms of volume the spaces occupied by the three main constituents are 87.1% inorganic, 1.4% organic and 11.5% water. The inorganic content decreases slightly from the enamel surface to the dentine–enamel junction, whereas the organic fraction has maxima at the surface and near the dentine with a minimum in the interior enamel[31]. Local variations in the density within enamel have been measured[43] by a density-gradient method after the systematic microdissection of sawn sections of enamel into particles by use of strong acid as a cutting tool[44]. The density of human enamel falls from approximately 3.01 at the surface to 2.89 near the dentine–enamel junction but local irregularities occur. Variations in the distribution of both major[45] and minor constituents such as carbonate[46] and fluoride[47] have been studied by development of sensitive techniques for the analysis of these dissected particles. Carbonate content rises from approximately 2.25% at the enamel surface to 3.9% near the dentine–enamel junction, being inversely related to density.

3. The inorganic phase

(a) Nature of the main constituent

(i) Chemical structure

As a result of early chemical analyses, enamel has long been known to consist mainly of calcium phosphates. Similarity of its hardness, density and optical properties to those of the mineral apatites suggested that the main inorganic component of enamel was hydroxyapatite, which can be represented by the formula $3 \, Ca_3(PO_4)_2 \cdot Ca(OH)_2$ and is in approximate agreement with the quantitative composition of enamel. In 1926 Gross[48] examined enamel by X-ray diffraction and found that it gave a pattern closely resembling that of apatite minerals. The positions of atoms in the apatite lattice were subsequently worked out by Mehmel[49] and Náray-Szabó[50].

In addition to the calcium, phosphate and hydroxyl ions required by the hydroxyapatite formula, the inorganic phase of enamel contains substantial amounts of carbonate, significant quantities of the "physiological" ions sodium, magnesium, potassium and chloride and much smaller amounts of many elements including fluoride and zinc. The content of the most abundant constituents is summarised in Table V. A great range of adventitious elements may therefore be found in enamel either (1) in the body of the crystallites, (2) adsorbed on the crystal surfaces, (3) in the electro-statistically held hydration shell of the crystallites or (4) associated with the organic components. Departure of the inorganic phase from the ideal hydroxyapatite composition is attributable to several effects (1) substitution in the crystal lattice e.g. strontium for calcium, carbonate for phosphate and fluoride for hydroxyl ions, (2) surface adsorption of such ions as sodium, magnesium, bicarbonate or citrate and the ability of the crystal to bind water and hydrated ions. In particular, values for the ratio Ca:P are usually, as in bone, rather less than the theoretical value of 2.15 (by weight) or 1.67 (atomic) for apatite. This has been variously explained as due to (1) adsorption of excess phosphate on crystal surfaces, (2) substitution of sodium and magnesium for calcium and (3) substitution of hydronium (H_3O^+) ions for two adjacent calcium ions in the crystal lattice. The relative contributions of these various factors which cause significant departure from stoichiometry not only in enamel but in bone and dentine are as yet unknown.

TABLE V

MAJOR INORGANIC CONSTITUENTS OF HUMAN ENAMEL

Values expressed as % by wt.

	Thermal neutron activation analysis[51] Sound	Chemical analysis[52]	
		Sound	Carious
Ca	37.4 ± 1.0	36.75 ± 0.17	35.95 ± 0.21
P	18.3 ± 2.2	17.41 ± 0.04	17.01 ± 0.06
CO_2	—	2.42 ± 0.02	1.56 ± 0.03
H_2O	—	2.02 ± 0.04	3.07 ± 0.05
Na	1.16 ± 0.40	(0.25 — 0.9)	—
Mg	0.36 ± 0.04	0.54 ± 0.01	0.40 ± 0.01
Cl	0.65 ± 0.30	(0.19 — 0.30)	—
K	—	(0.05 — 0.30)	—
F	—	0.012	0.058
Ca/P (wt.)	2.04	2.09 ± 0.02	2.08 ± 0.03

On treating enamel with acid, 2.5–3.5% by weight of carbon dioxide is released which can only be present as carbonate ions. This carbonate must be regarded as a major constituent but its relation to the hydroxyapatite lattice is uncertain. The three main possibilities are (1) that carbonate is present in a second solid phase, (*2*) that it is substituted for another ion in the apatite lattice, (*3*) that carbonate is adsorbed on the crystal surfaces. There is no evidence of a second crystalline phase, so if a second phase is present it must be amorphous. The most likely ion to be substituted by carbonate appears to be phosphate rather than hydroxyl[32]. However, the a-axes of enamel unit cells are 0.02 Å longer than in a standard hydroxyapatite (Table VI). The chloride content of enamel would account for an increase

TABLE VI

DIMENSIONS OF UNIT CELLS IN THE INORGANIC PHASE OF HUMAN DENTAL ENAMEL AND SOME SYNTHETIC CALCIUM APATITES[32]

	Dimensions in Å	
	a_0	c_0
Enamel	9.44$_2$	6.88$_3$
Hydroxyapatite	9.42$_2$	6.88$_3$
Chlorapatite	9.63$_4$	6.77$_8$
Fluorapatite	9.37$_3$	6.88$_2$

of 0.01 Å. However, the carbonate content would be expected to shorten the a-axis by 0.02 Å, judging from the effect of incorporation of carbonate on other well-crystallized carbonate apatites[32]. Thus there is a discrepancy of 0.03 Å between the observed and expected values for a_0. In a detailed examination of the infrared absorption spectra of carbonate-containing apatites, Elliott[53] concluded that only 5–10% of the total carbonate of enamel is present in the lattice and probably replaces hydroxyl ions, because the plane of these carbonate ions is parallel to the c-axis. Most of the carbonate ions had their plane approximately perpendicular to the c-axis and are assumed to be adsorbed on the crystal surfaces.

(ii) Physical characteristics

The extreme hardness of enamel in comparison with all other vertebrate materials is perhaps its most striking characteristic. This results from the intrinsic hardness of apatite (5 on Moh's scale, Knoop hardness 430 kg/mm^2)[32] and the close packing of the large crystallites with comparatively small intercrystalline spaces containing the aqueous and organic phase. The density[32] of hydroxyapatite is 3.1 g/ml while that of mature enamel from human permanent teeth lies mainly within the range 2.9–3.0.

The crystallites in enamel are considerably larger than those of collagenous mineralized tissues such as bone (Table VII) and dentine. Hippopotamus enamel and fish bone were chosen for comparison because their highly oriented crystallites enable accurate measurements to be made. Dimensions for other species, including human enamel[54], by electron microscopy, are in reasonable agreement with the values in the table which were obtained by calculation from X-ray diffraction data. The comparatively large size of the inorganic crystals is one of the most important characteristics of enamel and affects both its physical properties and chemical reactivity. The volume of individual enamel crystals is approximately 200 times greater than those in bone[55]. The area of crystallite surface is correspondingly smaller in enamel so that the inorganic phase is less reactive. Stability is enhanced by the close packing of the crystals, so that diffusion through the intercrystalline phase is a comparatively slow process. Glas[55] considers mature enamel as a colloidal fibre structure composed of apatite crystals and water, "contaminated" by organic as well as inorganic matter. Interchange between crystalline and intercrystalline phases may nevertheless be very important as flow of water occurs through intact enamel[56], amounting to approximately 4 $mm^3/cm^2/24$ h *in vitro*[57].

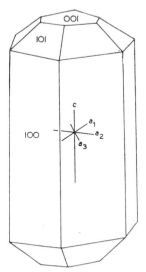

Fig. 4. Ideal crystal form of apatite showing the relation between the crystallographic axes a_1, a_2, a_3 and c and the faces of the hexagonal prism (100), pyramid (101) and base (001).

The detailed structure of the apatite crystal lattice is now widely accept-ed[58,59]. Apatite belongs to the hexagonal system, the unit cell having a sixfold c-axis and three equivalent a-axes perpendicular to the c-axis and at 120° to each other. The relation of these axes to the crystal faces is shown in Fig. 4 and the unit-cell dimensions are summarised in Table VI. The crystal habit is such that it is elongated almost parallel to the c-axis of the unit-cell in both enamel and bone (Table VII) although Hirai[60] claims there is a small divergence (approx. 2.2°) in enamel. The X-ray diffraction of enamel suggests that not only are its crystals larger than in bone but also more perfect. Crystallinity values (X) for mineral hydroxyapatite, 100; human enamel, 70; human dentine and bone, 35 and amorphous calcium carbonate phosphate, 17 have been reported by Trautz[32]. The difference between the values of X for enamel and bone crystals may be exaggerated by the presence of larger amounts of organic matter in bone causing scatter-ing and so reducing the value of X. Because of the sharp diffraction pattern produced by the large and perfect crystallites it is not possible to determine whether enamel contains amorphous calcium phosphate, but it seems prob-able that, if it does, the amount must be much less than in bone.

TABLE VII

AVERAGE CRYSTALLITE SIZE IN HIPPOPOTAMUS ENAMEL[76] AND FISH BONE[77]
Dimensions are in Å and are corrected for strain.

	Width perpendicular to c-axis	Length parallel to c-axis
Enamel, mean	410	1600
range	(370–450)	(1040–2700)
Bone	40–45	600–700

Because apatite is optically uniaxial it has different refractive indices in directions parallel and perpendicular to the c-axis. The difference between these refractive indices is the *intrinsic birefringence* which is characteristic of the substance. For apatite it is small and negative (-0.004) as a result of the high proportion of tetrahedral phosphate ions which produce a lower birefringence than planar ion groups such as carbonate. A second type of birefringence, *form birefringence*, is present in enamel and results from the pattern of oriented crystallites in relation to the inter-crystalline phase.

Texture in enamel arises from the arrangement of crystallites with respect to one another. As explained on p. 788 they are primarily grouped into rods or prisms, which in turn are approximately parallel to each other with some crossing over and mutual twisting. Crystal orientation has been determined within a single rod[61] and for a number of straight rods[62]. The average direction of the c-axes (and hence long axes) of the crystallites is approximately parallel to the rod axis but deviates in a direction towards the root of the tooth by an angle which increases from only 6° at the dentine–enamel junction to 23° near the enamel surface. The directions of individual crystallites vary from this mean direction, as already explained, defining the prism boundaries in mature enamel[21].

(iii) Mechanism of crystal growth

The composition, type, size and shape of crystal formed depends on the composition of the solution both as regards the concentration of inorganic ions and the presence of organic molecules which initiate or modify crystal growth. The phosphate:carbonate ratio has an important influence on the crystalline substance obtained as shown by *in vitro* experiments[32]. The presence of phosphate ions inhibits the formation of crystalline calcium carbonate from metastable solutions unless the ratio is below 1:300 for dilute solutions or 1:100 for more concentrated solutions. Below a ratio of 1:40 apatite tends to form[63]. Human serum ultrafiltrate has a ratio of approximately 1:22 and is undersaturated with respect to calcium carbonate but supersaturated to hydroxyapatite. However, artificially prepared solutions with similar concentrations of inorganic ions are stable indefinitely out of contact with solid calcium phosphates.

Brown[64] has suggested that enamel crystals grow as octacalcium phosphate $Ca_8H_2(PO_4)_6 \cdot 5 H_2O$ and are subsequently "hydrolysed" to hydroxyapatite, the entire process taking place in three stages. The first stage, which, as suggested in the previous paragraph, does not take place *in vitro* but occurs in mineralizing tissues in the presence of organic polymers, is as badly defined in enamel as it is in other mineralized tissues. It would appear to involve the formation of an "incipient seed", a grouping of ions sufficiently large to survive thermodynamically and subsequently grow into a crystal. The idea of the second stage is based on the initial rapid growth of ribbon-like crystals in young enamel, observed with the electron microscope. This is assumed to be due to a single unit-cell thickness of octacalcium phosphate growing in two dimensions only (*i.e.* in length and width but not in thickness).

The third stage is based on the subsequent and slower growth in thickness of crystallites, after the earlier rapid growth of ribbons along the prisms. It is assumed to occur in two steps in the first of which a layer one unit-cell in thickness is precipitated onto the 100 face of the crystal (see Fig. 4). The second step involves the hydrolysis of a unit-cell thickness of octacalcium phosphate to produce a layer of hydroxyapatite two unit-cells thick.

$$Ca_8H_2(PO_4)_6 \cdot 5\,H_2O + 2\,Ca^{2+} \rightleftharpoons 2\,Ca_5OH(PO_4)_3 + 4\,H^+ + 3\,H_2O$$

Whether growth in thickness occurs as hydroxyapatite or as octacalcium phosphate will depend on the relative rates of the two steps in the third stage.

A mechanism of this kind would explain the rapid initial growth of thin crystals along enamel prisms in the direction of retreating ameloblasts resulting in rapid preliminary mineralization of the young enamel. This is followed by a much slower growth in thickness during the "maturation" stage as the steady increase in mineral content shown by Deakins[24] occurs. It also explains the part played by fluoride in producing more rapid growth of highly ordered apatite crystals, because fluoride is known to catalyse the hydrolysis reaction. Further evidence to support this mechanism of crystal growth is provided by recent nuclear magnetic resonance studies of Casciani[42], which indicate the possible presence of $1-2\%$ octacalcium phosphate in enamel.

The low solubility of solid hydroxyapatite contrasted with the apparent difficulty of its precipitation from homogeneous metastable solutions still presents something of a paradox. The intrinsic instability of small aggregates of ions taking the form of the apatite lattice seems to be the core of the problem. Brown[64] has pointed out that at pH 7.4 and 37° octacalcium phosphate is approximately 10 times more soluble than hydroxyapatite. In a newly formed crystal of hydroxyapatite having surfaces of octacalcium phosphate there is the possibility of a "solubility gap" in the ionic product for calcium and phosphate between the high level necessary for further deposition of octacalcium phosphate and the substantially lower one at which solution of apatite would take place. Recently Squires[65] has made a new investigation into the operation of the Kelvin effect on the rate of precipitation of calcium phosphates from solution and has obtained evidence *in vitro* that solubility increases rapidly with reduction in size of aggregates.

(b) Minor inorganic constituents

The hydroxyapatite crystals in enamel can adsorb or exchange ions from the solution which surrounds them. Since enamel remains porous this process can take place during mineralization and even continue at a slow rate after tooth eruption. A great variety of elements, some at extremely low levels, is therefore found in association with the inorganic phase. Among the more minor constituents, Söremark and Samsahl[51] found the following levels in human enamel, results being expressed in parts per million determined by thermal-neutron activation analysis: zinc, 276 ± 106; strontium, 94 ± 22; bromine, 4.6 ± 1.1; tungsten, 0.24 ± 0.12; copper, 0.26 ± 0.11; manganese, 0.54 ± 0.08; and gold, 0.02 ± 0.01; while Söremark and Lundberg[66] obtained iron, 388 ± 109; rubidium, 4.9 ± 1.6; silver, 0.0049 ± 0.0012; chromium, 0.0027 ± 0.0016 and cobalt 0.00024 ± 0.00009.

Minor constituents are not necessarily evenly distributed through the thickness of the enamel. If they are strongly bound they will remain where they were first deposited and their position in relation to the enamel surface will indicate the direction from which they entered the enamel. The concentration of fluoride is highest at the outer surface and falls sharply at increasing distances from it within the enamel[31]. Since concentrations at all points increase with the fluoride content of the drinking water, this indicates a "filtering" effect of the porous enamel with fluoride replacing hydroxyl ions in the surfaces of hydroxyapatite crystals. The distribution of zinc, lead, iron and tin is similar to fluorine with highest concentrations near the enamel surface[31]. Strontium and copper have an even distribution through the thickness of enamel, while the more "physiological" ions carbonate, magnesium and sodium have highest concentrations near the dentine–enamel junction. This suggests the possibility that these ions are derived from tissue fluid which has passed through dentine. There is no evidence for the substitution of magnesium in the apatite lattice[32]. Sodium may replace calcium in the lattice, simultaneously with carbonate replacing phosphate[32].

4. The organic phase

(a) The protein content of young enamel

An amino acid analysis of a hydrolysate of whole enamel at the maturation stage from unerupted human deciduous teeth carried out in 1960, first showed the presence of substantial amounts of proteinaceous material having an unusual and characteristic composition (Table VIII) unlike that of any known protein[27]. These results were soon afterwards confirmed for developing bovine[67], pig[68,69] and human enamel from the permanent dentition[69-71]. The composition of the protein in young enamel had a number of striking features which were shown or approached in all three species. (1) A very high content of proline, which accounted for approximately a quarter of the total amino acid residues. (2) A high content of glutamic acid and frequently a high molar ratio relative to aspartic acid ($>4:1$). (3) A high content of histidine both absolute and relative to the other basic amino acids (molar ratio of histidine:lysine:arginine approximately $3:1:1$). (4) Tyrosine greater than phenylalanine. (5) High methionine compared with many proteins. (6) Cystine absent or present in very small amounts. (7) Hydroxylysine absent or present at a very low level (<3 residues per 1000). (8) Hydroxyproline absent or, if present, at a level less than 12 times the molar hydroxylysine content. (9) Glycine less than one tenth of total residues and often approximately one fifteenth.

The low glycine content and the absence or virtual absence of hydroxyproline show that young enamel protein is not a collagen. The occasional occurrence of small amounts of hydroxyproline may indicate contamination by collagen, a view supported by the finding of hydroxylysine in these samples. The amount of hydroxylysine present, though very small, exceeds what would be found in contaminating collagen, assuming the hydroxyproline to be entirely derived from this source. Hydroxylysine may therefore be a true constituent of young enamel protein[67] but is present at such a low level that it is near the limit of detection.

The composition of young enamel protein is also quite different from that of the keratin group. Cystine in enamel protein is substantially lower than in epidermis from skin or oral epithelium[72] which show the smallest amounts in any recognised keratins. The balance of basic amino acids in protein from young enamel with histidine greatly predominating over lysine and arginine is very different[73] from that in the epidermal keratins which also have a higher serine content than enamel protein.

Enamel protein not only appears to have no affinity with collagens and keratins, with which it was once compared, being an extracellular tissue protein of ectodermal origin, but it appears to differ in composition from all known proteins. Its very high content of proline remained a unique distinction until the discovery of a protein component of parotid saliva in which proline accounts for one third of the residues[74] but which does not have an exceptional histidine content. This parotid protein has another

TABLE VIII

AMINO ACID COMPOSITION OF TOTAL PROTEIN OF DEVELOPING ENAMEL IN MAMMALS

Values are given as residues of amino acid per 1000 total residues.

	Equidae	Suidae	Bovidae				Canis	Rodentia		Primates		
	Horse	Pig	Lamb	Llama	Goat	Ox	Dog	Rabbit	Rat	Squirrel monkey	Rhesus monkey	Man
Reference:	78	78	78	78	78	79	78	80	80	78	78	27
Cystine (half)	—	Tr.	—	—	8	0.22	—	1.5	1.6	Tr.	—	4
Hydroxyproline	—	—	Tr.	—	—	Tr.	—	—	—	—	Tr.	0
Aspartic acid	40	39	36	45	70	33	58	29	24	29	28	30.3
Threonine	30	39	31	33	38	27	32	21	32	38	42	38.1
Serine	63	50	49	46	95	49	58	41	74	60	50	62.5
Glutamic acid	153	152	159	144	173	159	153	98	178	151	156	142
Proline	221	218	230	241	196	253	221	253	240	245	264	251
Glycine	61	69	60	60	89	54	61	72	49	63	55	65.0
Alanine	20	24	24	21	34	22	32	145	41	20	21	20.3
Valine	36	35	38	52	27	38	41	33	36	40	42	39.6
Methionine	44	54	52	56	26	55	47	2	40	52	51	42.3
Isoleucine	38	38	34	30	24	34	36	12	37	33	25	32.7
Leucine	103	95	94	71	76	95	83	132	93	88	92	91.3
Tyrosine	58	51	49	56	27	47	43	33	32	58	52	53.4
Phenylalanine	25	25	23	26	21	26	24	19	21	20	16	23.4
Hydroxylysine	Tr.	—	Tr.	Tr.	Tr.	Tr.	—	—	—	—	—	0
Lysine	17	16	20	19	23	16	26	20	17	18	17	17.7
Histidine	56	71	68	66	54	73	69	64	73	68	66	64.5
Tryptophan	10	4	15	10	—	Pr.	Tr.	Pr.	Pr.	8	4	—
Arginine	18	20	19	18	19	19	18	21	11	10	16	23.3

Tr. = trace; Pr. = present; Tryptophan values are approximate.

characteristic in common with enamel in that it is a product of actively secreting ectodermal cells, which lends support to the view that enamel is an extracellular calcified secretion rather than a keratin system. Watson[75] has shown that enamel is separated from the cytoplasm of ameloblasts by their plasma membrane and is therefore extracellular.

For convenience the term "amelogenins" has been suggested[81,82] for the characteristic proteinaceous components, rich in proline and histidine, which predominate at an early stage of maturation and presumably have a role in enamel formation, but do not persist in mature enamel. The term "enamelins" has recently been used for the same components[83] but it would seem preferable to restrict this to proteins in mature enamel thereby distinguishing them from those which predominate at an early stage.

The amino acid composition of developing enamel from a wide range of mammalian species has recently been determined[78,80]. The results (see Table VIII) show that the picture of amino acids in amelogenins already described applies to a wide range of animals and possibly throughout the mammalia. A minor variation occurs in rabbit enamel which has diminished contents of glutamic acid and methionine and an unusually high alanine content compared with young enamel in other species. The histidine and proline contents in the rhesus monkey appear to be rather low but this may reflect a later stage of development in a particular specimen. Tryptophan at a level up to 15 residues per 1000 appears to be present in some species[78] and has been shown by autoradiography to be incorporated into young enamel[84]. Histochemical techniques suggest an abundance of sulphydryl groups in enamel protein[85] but analytical evidence does not support this (Table VIII).

(b) The protein content of mature enamel

Mature enamel contains a very low level of protein, corresponding to its high degree of mineralization. A value of 0.06% is reported for the total protein, peptide and amino acid content of enamel from erupted bovine incisors[34]. The outer enamel of human teeth has a similarly low level[86] of 0.05%, although total human enamel contains 0.3–0.4% of protein for molars and premolars and 0.2–0.25% for incisors and canines[86].

(i) Early investigations based on sedimentation of enamel

The low protein level of mature enamel together with its occurrence mechanically bonded to dentine and sometimes cementum, both of which contain approximately 20% of collagen, render the preparation of mature enamel protein technically difficult. The three earliest preparations of protein from mature human enamel for which substantially complete amino acid analyses are available[38,87,88] all appear to be contaminated with 40–50% of extraneous collagen, judged from their high contents of hydroxyproline and glycine. This contamination occurred despite attempted purification of powdered enamel from the dentine by sedimentation in a mixture of organic liquids of suitably high density[89], because the comparatively few "junction particles" which remained with the enamel contained as much weight of protein as that which was sparsely distributed in the much greater bulk of true enamel. On the basis of an incomplete analysis of mature enamel protein, which gave a molecular ratio of the basic amino acids histidine, lysine and arginine as 1:3.3:9.6, Block et al.[90] suggested it belonged to the "eukeratin" group, which was defined[91] as containing keratins similar to cattle horn, resistant to solution and to enzymes and having a molar ratio of basic amino acids of 1:4:12. The subsequent concurrence with this idea purely on the basis of the basic amino acid ratio[38,87,88], is clearly not justified[73], as this ratio varied widely (1:3.5–4.7:5.2–12.5) in the enamel samples analysed over a range which spans the accepted values for the collagen contaminant. Moreover there was a wide variation in other amino acids and the cystine content was too low for a keratin resembling horn.

(ii) Proteinaceous material in middle and outer enamel

Lofthouse[92–94], instead of separating enamel by sedimentation, dissected away dentine before pulverization and achieved a reduction in the level of contamination by collagen to 15% in a soluble fraction of human mature enamel protein (Table IX), although his insoluble fraction was heavily contaminated. The first substantially uncontaminated preparation of protein material from mature enamel was obtained in 1964 by Glimcher et al.[34] who employed a carefully controlled dissection technique on erupted bovine incisors. A thin layer of cementum, which covers the crowns of bovine teeth[95], was removed under the dissecting microscope after surface demineralization with EDTA for 2 days. The crowns were then cut into 400-μ thick sections and the middle third of the enamel selected by scoring a line parallel to the dentine–enamel junction and suitably far from it and then breaking off the enamel up to the line. This technique yielded enamel which was uncontaminated with collagen, either from the dentine or coronal cementum. The total protein of this purified enamel was obtained in the form of amino acids by direct solution of the enamel and hydrolysis in 6 N hydrochloric acid, with subsequent precipitation of calcium phosphate at pH 9 under conditions which did not result in adsorption of amino compounds by the precipitate. A second portion of enamel was demineralized

TABLE IX

AMINO ACID COMPOSITION OF VARIOUS PROTEIN PREPARATIONS FROM MATURE
ENAMEL

Values are given as residues of amino acid per 1000 total residues.

	Bovine					Human		
	Water-sol. degraded protein of wide distrbn.			soluble		Insol. protein of inner enamel	Acquired pellicle	
	whole enamel	non-diffusible						
		total	water-sol.					
Reference:	34	34	34	92,94	86	86	96	97
Cystine (half)	11	11	14	—	4	20	—	13
Hydroxyproline	0	0	0	15	8	2	—	—
Aspartic acid	94	88	84	73	54	79	70	71
Threonine	48	53	56	39	42	52	37	43
Serine	102	108	146	67	119	82	57	46
Glutamic acid	128	134	140	137	106	136	147	133
Proline	90	80	88	100	137	81	64	44
Glycine	195	161	154	177	193	62	128	81
Alanine	59	73	76	73	53	69	125	146
Valine	36	45	37	28	32	52	53	53
Methionine	5	1	9	—	34	22	—	12
Isoleucine	28	31	25	17	19	23	32	30
Leucine	67	71	54	44	66	111	59	64
Tyrosine	7	13	7	39	23	51	34	14
Phenylalanine	39	40	32	39	33	49	38	29
Hydroxylysine	1	1	1	13	4	6	—	—
Lysine	35	42	34	45	26	40	58	51
Histidine	26	16	21	11	19	27	35	19
Arginine	29	33	24	126	28	36	63	42
Ornithine								26
Muramic acid								21
Diaminopimelic acid								4
Hexosamines								53
Unknowns								3

in EDTA in dialysis bags and dialysed against water. The resulting preparations were hydrolysed for amino acid analysis directly (total, non-diffusible)
and after centrifugation (water-soluble).

Although 50–90% of the protein material was lost on dialysis the amino
acid compositions of the three preparations are quite similar (Table IX).
Hydroxyproline was absent, showing that collagen is not present in the
middle third of enamel, but approximately one residue of hydroxylysine per
thousand total residues was present, suggesting that mature enamel proteins
might contain a low level of this amino acid. The main features of their
composition are, however, high levels of glycine (though less than in collagen)
and serine, moderately high levels of aspartic and glutamic acids and low
levels of basic amino acids with lysine slightly predominating over histidine
and arginine. It is also clear from this study that the proteins from the middle
portion of the mature enamel are soluble in water, acid and EDTA and
are sufficiently broken down that a substantial proportion can pass dialysis
membranes.

Further work on the fractionation of organic constituents from the
enamel of 5000 erupted bovine incisors[98] showed that the most abundant
components were free amino acids and small peptides of molecular weight
less than 3500, consisting principally of glycine, serine and aspartic and
glutamic acids. Less than 10% of the organic content was of higher molecular
weight and consisted of two fractions separated on a polyacrylamide column.
The more abundant one had a molecular weight greater than 30000 and
was a glycoprotein with 90–95% carbohydrate, mainly glucose with some
galactose, mannose and fucose, with galactosamine predominating over
glucosamine. The smaller fraction of molecular weight less than 30000
contained 85–88% of carbohydrate in which galactose predominated.

Weidmann and Eyre[33,86,94] attempted to prepare collagen-free protein
from the enamel of mature human teeth which because of their small size
and convoluted shape offered greater difficulty to microdissection by hand
than the relatively large and flat bovine incisors and necessitated a different
technique. A surface layer 5μ thick was initially removed after a brief treatment with acid, the roots and cracks in the enamel were covered with nail
varnish to protect these regions from attack from the dilute (1:5) hydrochloric acid subsequently used to etch away the outer enamel. The protein
which dissolved in the acid solution was analysed after removal of calcium
phosphate by precipitation and desalting on a Dowex 50 WX4 column,
eluted with 0.1 M piperidine, not more than 10% of the protein being lost

in these steps. The average amino acid composition from analysis of 15 preparations is given in Table IX. The hydroxyproline content varied from 5 to 9 residues per 1000, a considerably higher content than in bovine enamel[34] and it was considered that collagen was a minor contaminant accounting for 5–9 % of the protein, rather than a true constituent. Hydroxylysine was rather higher in human than in bovine mature enamel and the value exceeded what would be expected to be contributed by the collagen contaminant. The general pattern of amino acids in outer human enamel is otherwise quite similar to that in the middle third of bovine enamel with the exception that aspartic acid is rather lower and proline higher in the human material. The solubility in acid, failure to be precipitated by trichloroacetic acid and electrophoretic behaviour[33] of proteinaceous material in mature human enamel suggests that it is also present in a degraded state, largely broken down to peptides and amino acids.

(iii) Acid-insoluble protein in inner human enamel

The degraded protein material found widely dispersed in enamel accounts for only one quarter to one eighth of the total protein content of human incisors and molars, respectively[86]. Protein is distributed unevenly in mature enamel[99] with greater concentrations near the dentine–enamel junction where distinctive histological features, the enamel spindles and tufts, occur. When intact human enamel is completely demineralized with trichloroacetic acid, hydrochloric acid or EDTA, there is a residue having the form partly of ribbons and floss and partly of amorphous deposits, which is protein in nature[35] and probably accounts for most of the total weight of protein in human molars and premolars. This material appears to be closely associated with the enamel nearest the dentine surface to which it loosely adheres after demineralization. Its occurrence there would account for the lower density of enamel in the innermost regions[43].

The ribbons, floss and amorphous material are intimately connected with one another and appear to be associated both with a membrane at the dentine–enamel junction and with organic lamellae which extend from the dentine to the enamel surface. Their origins and structural significance are still uncertain, but the features they show with the optical and scanning electron microscopes suggest that they are structurally associated with the large-scale, prismatic organisation of enamel and no underlying fibrous basis has been demonstrated. The amino acid composition is characteristic: an average from 6 preparations[86] is given in Table IX. The nature of the agent used for demineralization has little effect on the composition of the protein residue and a preparation rich in amorphous material differs little in composition from others consisting mainly of ribbons and floss[35]. Possibly the same kind of material fills "spaces" of different shapes. The composition of the insoluble residue is not totally unlike that of the more widely dispersed, degraded protein material but it is richer in leucine and cystine and poorer in glycine. In some respects there is a rather close resemblance to the low-sulphur keratin of human oral epithelium[72], which is a pseudo-keratin according to Block's classification[91], but this resemblance may be fortuitous.

(iv) Post-eruptive integument of enamel

During the later stages of enamel maturation the ameloblast layer degenerates and soon after eruption it is replaced on the enamel surface by a complex integument, acquired from sources other than the enamel itself. Meckel[100] distinguishes three layers, a sub-surface cuticle over which is a very thin surface cuticle covered in turn by a thicker pellicle. Over this pellicle there may be in places a plaque composed of a meshwork of bacteria and extracellular polysaccharides produced by them. Plaque is revealed by staining with basic fuchsin and may be removed by scrubbing with detergent leaving the three layers of the integument intact. These can be detached from the enamel after brief surface demineralization with dilute acids. Recent studies of their composition show the presence of some 50% of protein[101] having a similar composition to salivary glycoproteins[96] especially to submandibular mucin[101]. Leach *et al.*[96] considered that the acquired pellicle is substantially free from bacterial remnants and is mainly composed of specific salivary glycoproteins with their polypeptide chains intact and oligosaccharide groups somewhat altered. Armstrong, how-ever, has demonstrated the presence in human acquired pellicle of muramic acid[101] and diaminopimelic acid[97], substances which are common constituents of bacterial cell walls but not of animal proteins. He also showed that in the mean result from 8 analyses of pellicle (Table IX) almost all the amino acid values, including the exceptionally high alanine value, were inter-mediates between those for precipitated salivary mucin and bacterial cell wall. Since light and electron microscope examination of the acquired pellicle provides evidence of the presence of disintegrating bacteria it seems probable that they, like the salivary glycoproteins, contribute to its forma-tion, perhaps mainly in the outermost thick layer. The salivary constituents

References p. 831

may make their main and earliest contribution to the sub-surface and surface cuticles, in close proximity to the enamel proper.

(c) Heterogeneity of proteins in young enamel

(i) Solubility differences

The heterogeneity of the protein content of young enamel first appeared as differences in the composition of fractions dissolved by various solvents compared with that of the bulk protein. The water-[73] and neutral-soluble[79] fractions of enamel protein from unerupted teeth have higher contents of proline, glutamic acid and histidine than the total protein. These amino acids are characteristically high in the "amelogenin" fraction[81] which predominates at an early stage. During maturation there is an overall change in the relative proportions of amino acids in total enamel protein; proline, glutamic acid and histidine fall with time while glycine and serine increase in relative amounts[28] (Table III). The large increases in hydroxyproline and hydroxylysine in the last two columns probably resulted from contamination of the more mature enamel with extraneous collagen. The amelogenin fraction lost at an early stage[79] may be associated with the intraprismatic material[102] whereas the retained serine-rich fraction[34] is perhaps more related to prism boundaries.

Burgess[103] found that protein from young enamel was soluble in phosphoric acid but was precipitated by the addition of quite small concentrations of chloride, nitrate, sulphate or EDTA. Subsequently Burgess and Maclaren[28] separated 4 protein fractions, which differed somewhat in composition, by extraction and precipitation with various electrolytes.

(ii) Electrophoretic separations

Burgess[28,103] separated proteins from developing enamel electrophoretically using starch gels containing urea and formate, borate or glycine buffers. Closely similar results were obtained whether the starting material was prepared by dissolving protein from powdered enamel in 8 M urea or by precipitating it with chloride from acid or EDTA solutions used to demineralize enamel. It was therefore concluded that the multiple bands obtained were true components and not degraded artefacts. In most experiments, seven bands were obtained with component C dominant and J rather less abundant. Component C decreased markedly in relation to J as maturation proceeded thereby presumably accounting for the changes in composition.

High-voltage paper electrophoresis has been used to investigate soluble organic material from mature enamel[93]. This proved to be almost entirely of low molecular weight consisting of small peptides and free amino acids.

Polyacrylamide gels are capable of giving high resolution in protein separations. Initially difficulty was encountered in applying this method to enamel proteins. This was overcome by increasing the acrylamide concentration[104] to 18%. When bovine foetal enamel protein was subjected to electrophoresis at pH 4.2, one major component, 5 other principal components and 12 minor bands were separated. Gel electrophoresis has subsequently been used by several investigators as an analytical tool to examine components separated by other methods but its full potential on a preparative scale has not yet been exploited for enamel proteins (see p. 820).

Katz et al.[105] applied free-zone electrophoresis to the neutral-soluble proteins of embryonic bovine enamel. In alkaline solution they obtained one major boundary, representing 90% of the protein, and three minor boundaries, apparently representing non-interacting species. On dilution the pattern was unchanged except for increased asymmetry of the major boundary. In acid solution a more complex pattern appeared.

(iii) Column separations

Gel filtration on cross-linked dextrans has been widely explored for fractionation of soluble proteins from developing enamel. Burgess[28,103] applied a solution of chloride-precipitated soluble protein in dilute formic acid to a column of Sephadex G-50. The eluate monitored by its ultraviolet absorption was incompletely separated into two major overlapping peaks with indications of numerous smaller peaks. One major peak was shown by electrophoresis to consist mainly of component C and the other of a mixture of components H and J. Burrows[106] investigated the EDTA-soluble proteins in human enamel by gel filtration on Sephadex G-50 and detected three peaks in the eluate, only one of which contained hydroxyproline. It is not clear whether this applied only to mature or to developing enamel in addition. Weidmann and Hamm[93] used the ion-exchange materials DEAE cellulose DE-50, Amberlite CG-50 and DEAE Sephadex A-50 as well as Sephadex G-25 and obtained partial separations of components in an aqueous extract of mature human enamel.

The basic reason for incomplete separation of soluble proteins from young enamel in gel-filtration experiments emerged from the work of Mechanic *et al.*[107], who showed that a reversible aggregation–dissociation effect occurs between the components in aqueous solution, the position of equilibrium depending on the components present and their concentrations under prevailing conditions. When a given volume of a 1% solution of neutral-soluble protein from young enamel was applied to a Sephadex G-75 column, approximately 93% was eluted at the void volume (V_0). Since this material was excluded from the gel phase, its molecular weight exceeded 50000. There were two smaller peaks having V/V_0 values of 1.4 and 3.0, respectively, corresponding to material of lower molecular weight, which entered the pores of the gel. When the same volume of 1% protein solution was applied to a Sephadex G-100 column instead, less protein was eluted at the void volume and more entered the larger pores of the G-100 gel.

Five additional overlapping peaks, with V/V_0 values ranging from 1.4 to 3.6 were eluted. When the initial concentration of protein was reduced to 0.1% and the same volume of solution applied to the G-100 column, the peak at the void volume was even further reduced and a higher proportion of lower molecular weight material appeared in the slower moving peaks.

Dilution was considered to cause additional dissociation of the aggregates, which otherwise emerge at the void volume, with the production of smaller entities, which are retarded on the column. This explanation is probably correct as it is supported by ultracentrifugation studies, where a similar effect occurs[105]; otherwise it would be difficult to distinguish the effect from that produced by column overloading at the higher concentrations. On rechromatographing three subfractions of the high molecular weight material 97, 95 and 69% respectively were eluted at the void volume. Presumably this third subfraction had undergone further partial dissociation into compounds of lower molecular weight. Six original fractions which had been retarded on the Sephadex G-100 column at V/V_0 values of 1.2, 1.4, 1.8, 2.2, 2.5 and 2.8, respectively, when separately submitted to rechromatography, each gave a range of overlapping peaks in which the peak having the original V/V_0 value predominated, though several others were invariably present including unretarded high molecular weight material. Thus both aggregation and dissociation take place giving rise to larger and smaller aggregates, respectively, in peaks other than the original one. Some of the subsidiary peaks derived from a fraction, however, may have represented overlapping peaks originally present, since repeated chromatography gradually enriched the original peak up to a limiting value set by reaggregation.

All the rechromatographed peaks shared compositional features typical of amelogenins, namely high proline, glutamic acid and histidine contents. The type of protein high in glycine and serine which predominates later in enamel maturation did not appear in any of the fractions. Additional features involving especially high or low values for particular sets of amino acids were characteristic of individual peaks at specific V/V_0 values. These compositional features were fairly consistent for a given V/V_0 value in the second chromatogram, irrespective of the original V/V_0 value of the source material in the first chromatogram. The minimum and maximum values

TABLE X

SELECTED MINIMUM AND MAXIMUM VALUES FOR INDIVIDUAL AMINO ACIDS IN A SERIES OF FRACTIONS SEPARATED BY GEL FILTRATION AND COMPOSITIONS OF 3 FRACTIONS SEPARATED ON CM–SEPHADEX FROM DEVELOPING BOVINE ENAMEL PROTEIN

Values are given as residues of amino acid per 1000 total residues.

	Selected values from gel-filtration fractions[107]		Composition of fractions from CM–Sephadex chromatography[86]		
	Min.	Max.	A	B	C
Cystine (half)	—	17	—	—	—
Hydroxyproline	—	5.1	—	—	—
Aspartic acid	27	55	50	44	26
Threonine	13	38	35	33	26
Serine	20	100	90	96	42
Glutamic acid	75	236	131	134	204
Proline	180	356	182	176	250
Glycine	13	161	86	100	41
Alanine	3	38	35	23	22
Valine	19	58	29	33	44
Methionine	20	68	35	32	44
Isoleucine	18	41	34	32	39
Leucine	56	143	115	90	98
Tyrosine	—	130	42	67	35
Phenylalanine	15	40	31	31	20
Hydroxylysine	—	3.6	—	—	—
Lysine	4	33	31	30	11
Histidine	45	88	54	54	85
Arginine	4	45	20	24	12

for each amino acid, selected from analyses of a large number of fractions, are included in Table X.

(iv) The current state of knowledge

At the present time there is no clear picture of the way in which the protein components in newly laid down enamel are organised. Nevertheless it is apparent from changes in the content and composition of proteins in enamel during maturation, as well as by actual separation of protein fractions by various techniques, that enamel protein is heterogeneous, with components which differ considerably in molecular size and amino acid composition. However, not one of the ultimate molecular entities has so far been separated in a state of proven purity. The reason for this lies mainly in the readiness with which aggregation of the various molecular species takes place under separation conditions, as well as within the intact enamel itself. A picture of the extent and pattern of aggregation is revealed by recently published work.

Mechanic[83] found that addition of hydrogen-bond breakers (urea or guanidine thiocyanate) to solutions used for eluting neutral-soluble enamel proteins from Sephadex G-100 resulted in diminution of the peak emerging at the hold-up volume of the column and augmentation of the retarded peaks at V/V_0 values of 1.52, 1.96, 2.61 and 3.48, respectively, compared with other experiments where hydrogen-bond breakers were not added[107]. This effect has been confirmed by Nikiforouk and Gruca[108] and it presumably results from breakdown of aggregates into components of lower particle weight, which enter the gel pores. Increasing concentrations of urea led to a reduction in the size of the unretarded peak from 19.8 to 2.8% of the total protein, a smaller reduction in the peak at V/V_0 1.52 and increases in the 3 most retarded peaks[83]. The V/V_0 values of all peaks were constant throughout the experiments. Repeated chromatography of the unretarded peak gave rise to two peaks only, the original one together with the most retarded, the proportion of the latter being increased with higher urea concentrations. When the retarded peaks were rechromatographed each gave rise to 3 peaks only, the original peak, the unretarded aggregate and the most retarded peak. Here also, increasing the urea concentration caused further breakdown of the original peak. These effects were also observed by Nikiforouk and Gruca[108] on agarose columns and show the reversibility of the aggregation of retarded components into material of higher particle weight and also that this aggregation is partly dependent on hydrogen bond formation.

When the original enamel protein was chromatographed on the ion-exchange material DEAE cellulose in the presence of urea and a salt gradient, at least 19 components were partially separated, two of which were chromatographically homogeneous. Retarded material from the Sephadex G-100 column, which had appeared at the same elution volume in two successive experiments, proved heterogeneous on DEAE cellulose, the components at V/V_0 values of 1.52 and 2.61 containing 9 and 7 components, respectively[83]. In addition, all the gel-filtration peaks and the two chromatographically homogeneous peaks from the DEAE-cellulose column were found to be markedly heterogeneous when subjected to electrophoresis on polyacrylamide.

Fincham[104] chromatographed acid-soluble proteins from young enamel on DEAE cellulose and found that addition of urea resulted only in a sharpening of the protein peaks. Seven components were clearly distinguishable but there were indications of 18–20 peaks when stepwise changes were made in the potassium chloride concentration. With the possible exception of the original fastest and slowest peaks, all of the components isolated even when they appeared to be homogeneous on repeated chromatography under a variety of conditions, proved to be heterogeneous on subsequent polyacrylamide electrophoresis. The first group of proteins to be eluted from the DEAE-cellulose column were the amelogenins of high proline content. Fincham[109] suggested that since they show lipophilic properties, their aggregation may depend partly upon hydrophobic bonding, which would be unaffected by the presence of urea. A group of proteins eluted from the column considerably later contained electrophoretically fast-moving components which fluoresced, possibly as a result of their high tyrosine content.

Nikiforouk and Gruca[108] prepared, from rabbits, antisera to the total soluble proteins of bovine developing enamel. Only one of the four components separated from the enamel protein on a Sephadex G-100 column was antigenic to the serum. On polyacrylamide electrophoresis, 12 bands were separated from the original bovine amelogenin of which only the 6 nearest the origin proved to be antigenically reactive and identical, the faster moving bands of lower molecular weight apparently possessing no antibody-binding sites.

Weidmann and Eyre[86] subjected the total protein of developing bovine enamel to chromatography on the ion-exchange material CM–Sephadex C-50 eluted with a sodium phosphate gradient, containing urea. Approximately 8% of the total protein was eluted at the void volume and the remainder in three peaks designated A, B and C in order of increasing retardation and containing 12, 25 and 55% of the total protein, respectively. These three fractions were again subjected to chromatography on the CM-Sephadex when each gave rise to a single peak. The amino acid composition of the fractions after this purification step is given in Table X. Peaks A and B each gave a single band on polyacrylamide electrophoresis and were rather similar in composition except for tyrosine content. Peak C was electrophoretically heterogeneous and had a composition of the amelogenin type. Electrophoretic studies on the protein components present at the various stages of maturation showed that component C was lost earlier than A and B.

These recent studies give an impression of the complexity of the enamel-protein system. Failure to isolate and characterise any one chemically definable component, together with the wide range of composition shown by isolated fractions, suggests that covalently bound subunit components of enamel proteins are of many kinds and have fairly low molecular weights. There is probably a hierarchy of levels of aggregation with representative aggregates at each level being reproducibly definable in terms of behaviour in migration systems but less so as regards exact composition. Thus the very large aggregates (molecular weight $> 15 \cdot 10^6$)[108] which do not enter the pores in gel-filtration experiments reversibly dissociate into three[108] or four[83] fractions, which are retarded by the gels. These fractions are again separated into a number of components by ion-exchange chromatography on DEAE cellulose[83]. In turn, two of these components, which are chromatographically homogeneous can be resolved into even smaller units by electrophoresis on polyacrylamide gel. Thus 4 levels of aggregation are distinguishable using the somewhat arbitrary criteria of those methods which happen to have been used in attempts at separation.

Gel electrophoresis seems to be the most powerful method of separation since it produces most components and is able to resolve material which appears homogeneous on gel-filtration and ion-exchange columns. It is thus the most promising means for separating the fundamental units from which aggregates are built up by non-covalent interactions. Up to the present, electrophoretically separated components of enamel proteins have not been characterised by amino acid analysis, but newly developed apparatus for preparative gel electrophoresis and more sensitive analytical techniques render this feasible. A possible difficulty with gel electrophoresis is the production of more than one band per component as a result of dimer and trimer formation, through interaction with a buffer ion, *e.g.* acetate. This effect would be revealed in precise terms by subsequent amino acid analysis of the various bands.

A more serious difficulty is to ensure that all aggregates are completely broken down before or during electrophoresis. Three kinds of non-covalent interaction are likely to play a part in the aggregation of enamel proteins.

(*1*) *Electrostatic forces.* Amino acid analysis suggests that free side-chain carboxyl groups are present in amelogenins in numbers approximately equal to the amino and guanidino groups of basic amino acids[73]. The balance of charge is thus held by the imidazole groups of the numerous histidine residues and thus can be easily controlled by appropriate choice of the pH at which electrophoresis is carried out, *i.e.* in relation to the pK of the peptide bound imidazole group.

(*2*) *Hydrogen bonding.* This has been shown to occur extensively in enamel proteins as in many biological systems. It can be readily broken down by the addition of high concentrations of urea or even more effectively with guanidine or lithium salts.

(*3*) *Lipophilic bonding.* Formation of hydrophobic bonds has only recently been recognised as playing a major part in aggregate formation in enamel proteins. The problem is to reduce its effect in an aqueous environment. Since the hydrophobic bonds are mainly due to the preponderance of proline residues, addition of suitable water-soluble compounds which contain pyrrolidine groups[110] might be expected to reduce the interaction between proline residues attached to different chains.

Simultaneous reduction in aqueous solution of these 3 types of interaction and the application of appropriate separation techniques offer promise in the near future of solving the analytical problems which beset investigations into the nature of the protein complex present in young enamel.

(*d*) *The nature of enamel protein*

(*i*) *Amorphous character of the organic phase*

The existence of well-defined fibres in the organic phase of enamel has not been convincingly demonstrated. Early electron microscope observations suggested the formation of a fine fibrillar network[111,112] in the organic phase during maturation, while later it was suggested that organic sheaths surrounded individual apatite crystals[113]. Travis and Glimcher[22] observed in developing bovine enamel a series of dense lines 48 Å in thickness and having the appearance of filaments. The filaments which were parallel to one another and the prism axis could be resolved into two dense strands each 12 Å wide separated by a less electron-dense space 17 Å wide. The width of the bands is thus comparable with that of a single protein molecule. However, Boyde[114] has pointed out that the organic phase occupies all the space surrounding the crystals which, as they grow, compress the organic matter into the network of narrow channels between them. Hence on removal of the crystals by demineralization the organic matter has the appearance of a fibrillar network, as a result not of its basic structure but of the shape of the space it was constrained to occupy by crystal growth.

Fearnhead[115,116] prepared specimens of developing enamel by 5 different methods designed to bring about minimum alteration to the organic components. Electron microscopy showed no fibrils but only granules of approximately 50–70 Å diameter, infrared absorption showed absorption bands typical of proteins but no evidence of dichroism (which would result from orientation) and X-ray diffraction suggested an amorphous structure. It was concluded that there is no evidence for the insoluble organic component of developing enamel ever forming a highly ordered system of fibres and that the observations are more consistent with a gel structure.

Electron microscope and X-ray diffraction observations made on enamel which had been decalcified with basic chromic sulphate by Sundström and Zelander[117] gave no indications of fibrous structure and were consistent with a gel-like ground substance.

A gel can be prepared[12] from the main (presumably amelogenin) component of developing enamel by dissolving it in EDTA and adusting to pH 6.8–7.0 at 25° when a precipitate forms which can be separated by centrifugation. On storage of the precipitate at 4°, it forms a clear gel, which is reversibly transformed to a white opaque mass, presumably by separation of a second phase, on warming to 18°.

(ii) X-Ray diffraction

The X-ray diffraction behaviour of the organic phase of enamel from developing bovine teeth was investigated by Glimcher *et al.*[118] after demineralization with EDTA and drying the "membranes" obtained on siliconized glass. An oriented pattern was obtained with a strong meridional reflection at 4.65±0.05 Å and an equatorial one at 9.8–10.0 Å. This was interpreted as representing a cross-β structure[119] in which the polypeptide chains run transverse to the fibre axis. However, this configuration is rarely found in biological structures unless they have been exposed to denaturing conditions such as heat or acid[119-121]. This type of pattern was subsequently obtained by Hohling et al.[122], who interpreted it in terms of a parallel-β structure, and by Bonar et al.[123] for a specimen prepared from a gel of neutral-soluble enamel protein, by drawing it into fibres and allowing them to dry.

Pautard[124] also obtained an X-ray diffraction pattern which was consistent with the cross-β configuration from a specimen of human enamel which had been dehydrated with solvents and subjected to heat after demineralization. This treatment gave rise to doubt whether the cross-β configuration was present in the native state or whether it resulted from subsequent denaturation. In native specimens which had not been denatured no evidence for a β-pattern was found[125], though this pattern was readily produced by heating, acid treatment, or stretching[126]. Fearnhead[115,116] observed non-oriented spacings at 4.55 Å but no larger spacings and he concluded that his observations were inconsistent with α-helix, β or cross-β, structures for the major organic component of young enamel. Instead the X-ray diffraction data supported the infrared absorption and electron microscope observations indicating that the degree of order in the organic phase of developing enamel is low.

Those investigators who observed the cross-β pattern noted that it was accompanied by evidence for considerable unoriented material. It seems probable that the cross-β configuration is produced artificially by stretching during preparation of the specimen. Rudall[121] has pointed out that the same effect could be produced by forces resulting from the growth of apatite crystals, acting on the organic phase during maturation. Bonar et al.[127] consider that orientation of organic components may be lost during drying of material on glass so that the orientation of the morphological structural elements cannot be inferred from the diffraction data.

The X-ray diffraction characteristics of mature enamel which had been demineralized with EDTA and formic acid and fixed with formaldehyde were measured by Perdok and Gustafson[128]. This material, which presumably closely resembled the insoluble protein of Weatherell et al.[35] showed meridional spacings at 7 and 20 Å and equatorial reflections at 9.5 Å, consistent with the spacing of the protein chains. The 5.15 and 4.65 Å spacings characteristic of the α and β forms of keratin were missing and the material was consequently designated δ-(dental) keratin. The resemblance of these resistant materials in mature enamel to the pseudokeratins of Block has already been noted (p. 809).

References p. 831

(*iii*) *Mutability*

The organic phase of developing enamel is not a stable structure but a mobile and constantly changing one. This is most clearly shown by the manner in which radioactive isotopes of amino acids including histidine[129] and glycine[130] are incorporated, as shown by autoradiography. Incorporation through the ameloblasts is rapid but instead of the labelled material remaining as a sharp line (as occurs in dentine collagen) in just those regions of the organic phase which are undergoing synthesis at the time of administration, the label eventually spreads out through the entire thickness of enamel, including even those areas which were laid down before the labelled amino acid was given.

This mobility of the newly acquired amino acid suggests that the organic phase of enamel is much more mobile and perhaps more labile than the relatively static and permanent dentine fibres. Assuming that most of the isotope is rapidly acquired by the newly synthesized enamel proteins, these must be capable of some degree of diffusionable movement or flow, directed by concentration or pressure gradients through the thickness of the less mineralized parts of the newly secreted enamel. Since crystal growth occurs rapidly in those regions, the protein may be involved in the transport of inorganic ions from the surface of the enamel to the growing crystals at considerable depths within it. Histidine is sometimes implicated in metal binding by proteins and alternate changes from charged to uncharged state may provide a mechanism whereby it can alternately take up and release ions within or near the ameloblasts and at the apatite crystal surfaces, respectively. The greater degree of labelling by histidine[129] in enamel than in dentine and the relatively weaker labelling of enamel by radioactive glycine are parallel to the amino acid composition of the total enamel protein and dentine collagen respectively, suggesting that the isotopes are present in the enamel in combination with protein rather than as free amino acids. The longer retention of that part of the protein which is labelled with glycine is also in agreement with analytical studies of proteins present at various stages of maturation[28,95].

(*iv*) *The spatial pattern of maturation*

When enamel is first secreted by ameloblasts, it is slightly mineralized as indicated[131] by the density of approximately 1.5 and the presence of thin apatite crystallites close to the plasma membrane of the ameloblasts[54]. Maturation involves the growth of these crystals and the consequent diminution of the aqueous organic phase which surrounds them. During the course of maturation, all parts of the enamel are not at the same stage but there is a definite spacial pattern with some areas following after others in reaching any given degree of maturation. In general the pattern of maturation approximately follows the order in which the enamel was secreted by the ameloblasts.

The pattern of maturation can be readily traced by microradiography of thin sections of developing enamel, where the progress of a radio-opaque zone can be followed in teeth at different stages until it has spread throughout the enamel. In human teeth this zone begins at the apex of the dentine–enamel junction under what will become the cusp of the tooth, and spreads rapidly as a very narrow band along the dentine–enamel junction and simultaneously through the thickness of the enamel at the cusp until it reaches the outer surface of the enamel[132]. The zone of high opacity then spreads progressively towards the root of the tooth, with that portion immediately adjoining the dentine–enamel junction considerably, and that part near the enamel surface slightly ahead of that in the mid enamel, with the result that the front of high mineralization is concave. There appears to be a somewhat sharp line dividing the enamel recently laid down from the more mature zone, though maturation probably occurs continuously in both zones. In molar teeth, which have several cusps, maturation proceeds separately from each cusp with the mature zone eventually meeting in the fissures between the cusps.

Enamel which has been partially demineralized[133] with acetic acid buffer pH 3.5, when sectioned and stained with Papanicolaou EA 65 stain, gives a pattern closely similar to that shown by microradiography[134]. The newly laid down enamel stains red with the eosin component of the stain, which also selectively stains the enamel protein, after it has been isolated by complete demineralization with EDTA or urea extraction. The more mature zone in the sections is stained green by the light green SF component, which also selectively stains synthetic hydroxyapatite. There is thus a spatial correspondence between rapid increase of inorganic content shown by the radio-opacity and rapid loss of protein in maturing enamel, the change sweeping across the enamel in the pattern already indicated for human teeth. In rodent incisors, which grow continuously, there is a short concave junction to the green zone across the width of the enamel, the growing end staining entirely red and the incisal end green.

The sharp line in the microradiographs and Papanicolaou-stained sections would seem to indicate some sudden transition in the course of the maturation. Prior to this a steady acquisition of inorganic ions and crystal growth could take place, possibly with gradual loss of protein and water until a critical situation is reached when the system becomes unstable and sudden expulsion of most of the protein occurs. This was first suggested by Starkey[135], based on observations of the uptake of labelled calcium and strontium in the rabbit. A sudden shrinkage resulting in local distortion and reduction in volume may also take place. The idea of a critical stage is still highly speculative being based upon findings from different techniques the coincidence of which may be fortuitous. In disputing its existence, Boyde[136] rightly points out that, if such a stage occurs, there should also be an abrupt increase in mineral content, whereas electron probe analysis of calcium across developing enamel shows a smooth and almost linear increase, without discontinuity from the developing surface to completely mineralized sites[137].

(v) Mechanism of protein removal

The pattern of progressive mineralization in maturing human enamel is such that proteins are first removed from the enamel in the region of the dentine–enamel junction, which is relatively remote from the ameloblasts. This region consequently becomes highly mineralized at an early stage and so forms a rather impermeable barrier between the cell processes of the dentine and the mid-layers of enamel which are next to become heavily mineralized with consequent loss of protein. The pattern of protein loss is therefore such that it would not readily result from the action of a soluble product of either group of cells diffusing into the enamel when maturation is occurring.

For this reason it appears that, once enamel is secreted by ameloblasts and is separated from them by the considerable distance (compared with cellular dimensions) represented by the thickness of enamel, it is not under direct cellular control. Mechanisms for crystal growth and protein removal would therefore appear to be built into the structure of the secreted enamel. The overall structure of the enamel system appears to be such that, given some simple motive influence, such as a constant influx of calcium and phosphate ions, it is pre-set to go through a series of physico-chemical changes which result in a very highly mineralized condition.

It has been assumed by most investigators that degraded enamel proteins leave the enamel *via* ameloblasts at a late stage of maturation. Reith[138,139] has demonstrated by electron microscopy that the material is removed through ameloblasts and it has been shown that these cells possess an extensive lysosomal system[140,141], which contains acid phosphatase and other hydrolytic enzymes. The lysosomes probably contain proteolytic enzymes capable of breaking down enamel protein, since Sakamoto and Sasaki[142] have shown that the bovine enamel organ possesses catheptic activity which increases on a cellular basis as maturation proceeds. The most likely site for the action of lysosomal enzymes on enamel proteins is, however, not yet known.

The concept that developing enamel consists of a two-phase system with apatite crystals in a continuous non-fibrillar relatively structureless phase, consisting of proteins and water, provides an explanation for both the mobility of enamel protein and the route it eventually takes while passing out of the enamel. The moist organic phase takes the form of a series of connecting channels, continuous throughout enamel. The channels will be narrow in the most mature enamel nearest the dentine, because here the apatite crystals have grown to a larger size and so approach each other more closely. In the regions of less mature enamel, situated between the mature zone and the enamel surface, the channels will be wider because the apatite crystals are smaller. Thus proteins leaving the maturing zone have a pathway of relatively wide channels through which they can pass towards the ameloblasts on the surface, where they leave the enamel. This condition for effective removal will continue to apply as the more mature zone spreads across the enamel towards the surface, in the pattern described on p. 826.

Removal of water and protein closely correspond with increasing mineral content and it has been suggested[73] that changes directly associated with the growth of apatite crystals are responsible for the mechanism of removal. If the protein phase is a viscous sol, the increase in pressure and reduction in volume available for it, resulting from crystal growth, would automatically result in a squeezing out through the channels towards the surface. There would be a similar effect if the protein were in the form of a gel, although it seems necessary that the gel should have thixotropic properties[73] to enable it to flow most readily from regions where the pressure is greatest.

The discovery that amelogenins exist as a hierarchy of aggregates, which readily dissociate and reaggregate according to the prevailing conditions in solution, may account for their behaviour in enamel maturation. If dis-aggregation occurs in the more mature regions where crystal surfaces are growing towards one another, this would reduce the viscosity of the protein solution or bring about liquefaction of a gel and so facilitate flow. When the solution had moved away from the highly mineralized site and the conditions resulting in dissociation of aggregates no longer applied, reaggregation could occur resulting in a local increase in viscosity or gelation. A reversible process of this kind could occur repeatedly as mineralization proceeded, permitting complete mineralization of the older enamel while providing physical stabilization of the more newly formed parts. The mechanism of such a reversible aggregation process is not yet known but the concentration of inorganic ions in the immediate neighbourhood may have an important effect. Calcium and phosphate concentrations in solution would be expected to be greater in the younger enamel than in more mature regions. The aggregation phenomenon may also be linked to the transport mechanism through electrostatic interaction between small inorganic ions and side-chain groups of amelogenins. The high proportion of histidine in amelogenins may be instrumental in determining the state of aggregation since the net charge on the protein will alter rapidly with small pH changes in the region of pH 7. This is a possible alternative explanation of localised changes in the aggregation state of proteins in maturing enamel.

Other explanations of the mechanism of protein removal depend upon breakdown of covalent bonds. The suggestion[73] that the high proportion of proline increases the rigidity of the amelogenin molecule and so facilitates transmission of applied force to break specific weak linkages appears less plausible than the reversible aggregation concept which also is partly dependent on a high proline content for its operation[143].

The enzymatic breakdown of covalent bonds has already been mentioned as possibly occurring in ameloblasts when enamel proteins return through them. Katchburian and Holt[141] have suggested that lysosomes are secreted along with the enamel proteins, although this is not easily proved experimentally. If this is so and the protein hydrolases contained in the lysosomes play a part in protein removal, the pattern of breakdown by these enzymes must be consistent with the known pattern of mineralization and protein loss (p. 795). This suggests that the lysosomes remain intact in those parts of the enamel where maturation is yet incomplete but break down and release their contents where mineralization is reaching a high level. This would provide a mechanism of protein removal corresponding to the dissociation of aggregates except that it would involve breaking covalent bonds and would be irreversible. The aggregation phenomenon appears more attractive in accounting for the mutability and transport mechanisms of enamel protein. It is also possible that both processes occur, the reversible one in the less mineralized zones and the enzymatic one where mineralization is nearing completion.

(vi) Relation between the organic and inorganic phases
Unfortunately little is known concerning the relation between the organic and inorganic phases of enamel although this is very important in relation to the mechanism of crystal nucleation in newly secreted enamel and to the integrity of the fully mature product in erupted teeth. These aspects are conveniently treated separately.

The idea of an "organic matrix" in mineralized tissues as a pattern of ionic groups which attracts inorganic ions and stabilizes them in the appropriate relative positions to form seeds of apatite crystals was developed for bone[147]. The abundant organic constituent, collagen, has been most widely held to act as a matrix in bone[148], although its participation has never been finally proved and the epitactic centre has never been identified[82]. Recently attention has been turned from the matrix hypothesis to the direct production of apatite seeds by the cells of collagenous tissues.

In enamel there is less definite evidence for a protein nucleator mainly because very little protein remains in mature enamel. It is therefore necessary to postulate either that the epitactic protein must be a very minor constituent or else that there is a mechanism for detaching apatite crystals from their nucleating agent. Since enamel crystals are large and grow greatly in length they are consequently few in numbers which is compatible with the nucleator being a minor constituent. Glimcher *et al.*[144] have shown that both developing and mature enamel contain small amounts of both insoluble and soluble collagen. It is not clear, however, whether this is an impurity. Alternatively, first formed crystals of apatite in enamel may be nucleated by dentine collagen at the dentine–enamel junction where both formation and mineralization begins.

Glimcher and Krane[145] have provisionally shown that a substantial proportion of the serine residues in both developing and mature enamel proteins is phosphorylated. They isolated serine phosphate from partial hydrolysates of the proteins and identified it by its electrophoretic mobility. They suggested that the covalently bound phosphate groups are very reactive, especially towards calcium ions, and may thus act as bridges, linking protein to inorganic crystals; further orderly arrays of phosphate groups might be responsible for the nucleation of apatite crystals. The mechanism of nucleation in enamel is still very uncertain, however, and the possibility of direct participation of ameloblasts also remains.

It has been widely held that the small amount of organic matter which remains in mature enamel is tightly bound to the crystals and that the maintenance of this relationship is essential for the preservation of the integrity of enamel when threatened by disease processes. Darling[146] considers that the initial attack of dental caries is selective in its path through different histological elements of enamel. He produced evidence that soluble organic elements are dissolved at an early stage followed by progressive demineralization of the apatite crystals. Recent analytical investigations of the sparse organic components of outer and mid enamel[33,34] show that they are soluble peptides of acidic character, being rich in aspartic and glutamic acids and serine which is in part phosphorylated. Their retention on the apatite surface is probably fortuitous, resulting from electrostatic attraction of these negatively charged organic ions. They are the surviving representatives of the great bulk of enamel proteins and peptides which leave enamel during the maturation process. Their low molecular weight and solubility throws doubt on whether they have a highly specific role in protecting apatite surfaces within enamel. The mechanism of caries within enamel thus appears to be concerned mainly with diffusion of ions within the aqueous phase of mature enamel and the stability of apatite crystal surfaces under a range of local conditions.

REFERENCES

1   M. L. WATSON, *J. Biophys. Biochem. Cytol.*, 7 (1959) 489.
2   R. W. FEARNHEAD, *Arch. Oral Biol.*, 4, Spec. Suppl., (1961) 24.
3   D. F. G. POOLE AND M. S. GILLETT, *J. Dental Res.*, 48 (1969) 1119.
4   T. ØRVIG, in A. E. W. MILES (Ed.), *Structural and Chemical Organization of Teeth*, Vol. 1, Academic Press, New York, 1967, p. 45.
5   P. G. KELLY, P. T. P. OLIVER AND F. G. E. PAUTARD, *Proceedings of the 2nd European Symposium on Calcified Tissues*, Univ. of Liège, 1965.
6   P. T. LEVINE, M. J. GLIMCHER, J. M. SEYER, J. I. HUDDLESTON AND J. W. HEIN, *Science*, 154 (1966) 1192.
7   H. M. LEICESTER, *The Biochemistry of Teeth*, Mosby, St. Louis, 1949.
8   M. V. STACK AND R. W. FEARNHEAD (Eds.), *Tooth Enamel*, John Wright, Bristol, 1965.
9   A. E. W. MILES (Ed.), *Structural and Chemical Organization of Teeth*, Academic Press, New York, 1967.
10   M. V. STACK, *Ann. N. Y. Acad. Sci.*, 60 (1955) 585.
11   S. L. ROWLES, *Sci. Basis Med. Ann. Revs.*, (1962) 259.
12   G. NIKIFORUK AND R. F. SOGNNAES, *Clin. Orthopedics*, 47 (1966) 229.
13   S. M. WEIDMANN, *Dental Enamel: Rock or Tissue?*, Inaugural Lecture, Leeds Univ. Press, 1967.
14   M. V. STACK, *Odontol. Rev.*, 17 (1967) 101.
15   J.-G. HELMCKE, in A. E. W. MILES (Ed.), *Structural and Chemical Organization of Teeth*, Vol. 2, Academic Press, New York, 1967, p. 135.
16   M. U. NYLEN, *Intern. Dental J.*, 17 (1967) 719.
17   R. M. FRANK AND J. NALBANDIAN, in A. E. W. MILES (Ed.), *Structural and Chemical Organization of Teeth*, Vol. 1, Academic Press, New York, 1967, p. 399.
18   A. I. DARLING, K. V. MORTIMER, D. F. G. POOLE AND W. D. OLLIS, *Arch. Oral Biol.*, 5 (1961) 251.
19   A. BOYDE, in M. V. STACK AND R. W. FEARNHEAD (Eds.), *Tooth Enamel*, John Wright, Bristol, 1965, pp. 163 and 192.
20   D. F. G. POOLE, in A. E. W. MILES (Ed.), *Structural and Chemical Organization of Teeth*, Vol. 1, Academic Press, New York, 1967, p. 111.
21   A. H. MECKEL, W. J. GRIEBSTEIN AND R. J. NEAL, in M. V. STACK AND R. W. FEARNHEAD (Eds.), *Tooth Enamel*, John Wright, Bristol, 1965, p. 160.
22   D. F. TRAVIS AND M. J. GLIMCHER, *J. Cell Biol.*, 23 (1964) 447.
23   D. F. G. POOLE AND M. V. STACK, in M. V. STACK AND R. W. FEARNHEAD (Eds.), *Tooth Enamel*, John Wright, Bristol, 1965, p. 172.
24   M. DEAKINS, *J. Dental Res.*, 21 (1942) 429.
25   M. V. STACK, *J. Bone Joint Surg.*, 42B (1960) 853.
26   M. V. STACK, in A. E. W. MILES (Ed.), *Structural and Chemical Organization of Teeth*, Vol. 2, Academic Press, New York, 1967, p. 317.
27   J. E. EASTOE, *Nature*, 187 (1960) 411.
28   R. C. BURGESS AND C. M. MACLAREN, in M. V. STACK AND R. W. FEARNHEAD (Eds.), *Tooth Enamel*, John Wright, Bristol, 1965, p. 74.
29   J. E. EASTOE, *J. Dental Res.*, 46 (1967) 122.
30   I. M. SHAPIRO, R. E. WUTHIER AND J. T. IRVING, *Arch. Oral Biol.*, 11 (1966) 501.
31   F. BRUDEVOLD AND R. SÖREMARK, in A. E. W. MILES (Ed.), *Structural and Chemical Organization of Teeth*, Vol. 2, Academic Press, New York, 1967, p. 247.
32   O. R. TRAUTZ, in A. E. W. MILES (Ed.), *Structural and Chemical Organization of Teeth*, Vol. 2, Academic Press, New York, 1967, p. 165.
33   D. R. EYRE, *Ph. D. Thesis*, Leeds Univ., 1969.

34 M. J. GLIMCHER, U. A. FRIBERG AND P. T. LEVINE, *Biochem. J.*, 93 (1964) 202.
35 J. A. WEATHERELL, S. M. WEIDMANN AND D. R. EYRE, *Caries Res.*, 2 (1968) 281.
36 R. C. BURGESS, G. NIKIFORUK AND C. MACLAREN, *Arch. Oral Biol.*, 3 (1960) 8.
37 R. E. S. PROUT AND E. R. SHUTT, *Arch. Oral Biol.*, in the press.
38 M. V. STACK, *J. Am. Dental Assoc.*, 48 (1954) 297.
39 F. BRUDEVOLD, L. T. STEADMAN AND F. A. SMITH, *Ann. N. Y. Acad. Sci.*, 85 (1960) 110.
40 G. W. BURNETT AND J. A. ZENEWITZ, *J. Dental Res.*, 37 (1958) 581.
41 W. H. EMERSON, *J. Dental Res.*, 39 (1960) 864.
42 F. S. CASCIANI, in R. W. FEARNHEAD AND M. V. STACK (Eds.), *Tooth Enamel II*, John Wright, Bristol, 1971.
43 J. A. WEATHERELL, S. M. WEIDMANN AND S. M. HAMM, *Caries Res.*, 1 (1967) 42.
44 J. A. WEATHERELL, S. M. WEIDMANN AND S. M. HAMM, *Arch. Oral Biol.* 11 (1966) 107.
45 J. A. WEATHERELL, C. ROBINSON AND A. S. HALLSWORTH, in R. W. FEARNHEAD AND M. V. STACK (Eds.), *Tooth Enamel II*, John Wright, Bristol, 1971.
46 J. A. WEATHERELL, C. ROBINSON AND C. R. HILLER, *Caries Res.*, 2 (1968) 1.
47 A. S. HALLSWORTH AND J. A. WEATHERELL, *Caries Res.*, 3 (1969) 109.
48 R. GROSS, *Die kristalline Struktur von Dentin und Zahnschmelz*, Festschr. Zahnarztl. Inst. Univ. Greifswald, Berlin, 1926.
49 M. MEHMEL, *Z. Krist.*, 75 (1930) 323.
50 S. NÁRAY-SZÁRBO, *Z. Krist.*, 75 (1930) 387.
51 R. SÖREMARK AND K. SAMSAHL, *Arch. Oral Biol.*, 6 (1961) 275.
52 E. JOHANSEN, in M. V. STACK AND R. W. FEARNHEAD (Eds.), *Tooth Enamel*, John Wright, Bristol, 1965, p. 177.
53 J. C. ELLIOTT, in M. V. STACK AND R. W. FEARNHEAD (Eds.), *Tooth Enamel*, John Wright, Bristol, 1965, p. 20.
54 E. RÖNNHOLM, *J. Ultrastruct. Res.*, 6 (1962) 249, 268, 288.
55 J. E. GLAS, *Studies on the Ultrastructure of Calcified Tissues*, Tryclan Balder, Stockholm, 1962.
56 D. F. G. POOLE, P. W. TAILBY AND D. C. BERRY, *Arch. Oral Biol.*, 8 (1963) 771.
57 G. BERGMAN AND B. SILJESTRAND, *Arch. Oral Biol.*, 8 (1963) 37.
58 C. A. BEEVERS AND D. B. MACINTYRE, *Mineral. Mag.*, 27 (1946) 254.
59 A. S. POSNER, A. PERLOFF AND A. F. DIORIO, *Acta Cryst.*, 11 (1958) 309.
60 G. HIRAI, in M. V. STACK AND R. W. FEARNHEAD (Eds.), *Tooth Enamel*, John Wright, Bristol, 1965, p. 182.
61 D. CARLSTRÖM, *Advan. Biol. Med. Phys.*, 7 (1960) 77.
62 J. E. GLAS, *Arch. Oral Biol.*, 7 (1962) 91.
63 B. N. BACHRA, O. R. TRAUTZ AND S. L. SIMON, *Arch. Biochem. Biophys.*, 103 (1963) 124.
64 W. E. BROWN, in M. V. STACK AND R. W. FEARNHEAD (Eds.), *Tooth Enamel*, John Wright, Bristol, 1965, p. 11.
65 F. C. SMALES, in R. W. FEARNHEAD AND M. V. STACK (Eds.), *Tooth Enamel II*, John Wright, Bristol, 1971.
66 R. SÖREMARK AND M. LUNDBERG, *Acta Odontol. Scand.*, 22 (1964) 255.
67 M. J. GLIMCHER, G. MECHANIC, L. C. BONAR AND E. J. DANIEL, *J. Biol. Chem.*, 236 (1961) 3210.
68 K. A. PIEZ, *Science*, 134 (1961) 841.
69 K. A. PIEZ, in E. O. BUTCHER AND R. F. SOGNNAES (Eds.) *Fundamentals of Keratinization*, American Association for the Advancement of Science. Washington D. C., 1962, p. 173.
70 K. A. PIEZ AND R. C. LIKINS, in R. F. SOGNNAES (Ed.), *Calcification in Biological Systems*, American Association for the Advancement of Science, Washington D. C., 1960, p. 411.
71 K. A. PIEZ, *J. Dental Res.*, 39 (1960) 712.

72 J. E. EASTOE, *Arch. Oral Biol.*, 8 (1963) 449.
73 J. E. EASTOE, *Arch. Oral Biol.*, 8 (1963) 633.
74 I. D. MANDEL, *J. Dental Res.*, 45 (1966) 634.
75 M. L. WATSON, in E. O. BUTCHER AND R. F. SOGNNAES (Eds.), *Fundamentals of Keratinization*, American Association for the Advancement of Science, Washington D. C., 1962, Chap. 10.
76 J. E. GLAS AND K. Å. OMNELL, *J. Ultrastruct. Res.*, 3 (1960) 334.
77 D. CARLSTRÖM AND J. E. GLAS, *Biochim. Biophys. Acta*, 35 (1959) 46.
78 P. T. LEVINE, J. SEYER, J. HUDDLESTON AND M. J. GLIMCHER, *Arch. Oral Biol.*, 12 (1967) 407.
79 M. J. GLIMCHER, G. L. MECHANIC AND U. A. FRIBERG, *Biochem. J.*, 93 (1964) 198.
80 P. T. LEVINE AND M. J. GLIMCHER, *Arch. Oral Biol.*, 10 (1965) 753.
81 J. E. EASTOE, *Advan. Fluorine Res. Dental Caries Prevent.*, 3 (1965) 5.
82 J. E. EASTOE, *Calc. Tiss. Res.*, 2 (1968) 1.
83 G. MECHANIC, in R. W. FEARNHEAD AND M. V. STACK (Eds.), *Tooth Enamel II*, John Wright, Bristol, 1971.
84 W. R. COTTON AND S. M. HEFFEREN, *Arch. Oral Biol.*, 11 (1966) 1027.
85 D. BERARD, *Rev. Fac. Odontology Sao Paulo*, 3 (1965) 21.
86 S. M. WEIDMANN AND D. R. EYRE, in R. W. FEARNHEAD AND M. V. STACK (Eds.), *Tooth Enamel II*, John Wright, Bristol, 1971.
87 W. C. HESS, C. Y. LEE AND B. A. NEIDIG, *J. Dental Res.*, 32 (1953) 582.
88 G. C. BATTISTONE AND G. W. BURNETT, *J. Dental Res.*, 35 (1956) 260.
89 R. S. MANLY AND H. C. HODGE, *J. Dental Res.*, 18 (1939) 133.
90 R. J. BLOCK, M. K. HORWITT AND D. BOLLING, *J. Dental Res.*, 28 (1949) 518.
91 R. J. BLOCK, *Cold Spring Harbor Symp. Quant. Biol.*, 6 (1938) 79.
92 R. W. LOFTHOUSE, *M. Sc. Thesis*, Leeds Univ., 1961.
93 S. M. WEIDMANN AND S. M. HAMM, in M. V. STACK AND R. W. FEARNHEAD (Eds.), *Tooth Enamel*, John Wright, Bristol, 1965, p. 83.
94 S. M. WEIDMANN AND D. R. EYRE, *Caries Res.*, 1 (1967) 349.
95 M. J. GLIMCHER, U. A. FRIBERG AND P. T. LEVINE, *J. Ultrastruct. Res.*, 10 (1964) 76.
96 S. A. LEACH, P. CRICHLEY, A. B. KOLENDO AND C. A. SAXTON, *Caries Res.*, 1 (1967) 104.
97 W. G. ARMSTRONG AND A. F. HAYWARD, *Caries Res.*, 2 (1968) 294.
98 M. J. GLIMCHER AND P. T. LEVINE, *Biochem. J.*, 98 (1966) 742.
99 L. W. WACHTEL, *J. Dental Res.*, 38 (1959) 3.
100 A. M. MECKEL, *Arch. Oral Biol.*, 10 (1965) 585.
101 W. G. ARMSTRONG, *Caries Res.*, 1 (1967) 89.
102 M. J. GLIMCHER, D. F. TRAVIS, U. A. FRIBERG AND G. L. MECHANIC, *J. Ultrastruct. Res.*, 10 (1964) 362.
103 R. C. BURGESS, *J. Can. Dental Assoc.*, 29 (1963) 24.
104 A. G. FINCHAM, *Calc. Tiss. Res.*, 2 (1968) 353.
105 E. P. KATZ, G. L. MECHANIC AND M. J. GLIMCHER, *Biochim. Biophys. Acta*, 107 (1965) 471.
106 L. R. BURROWS, in M. V. STACK AND R. W. FEARNHEAD (Eds.), *Tooth Enamel*, John Wright, Bristol, 1965, p. 59.
107 G. L. MECHANIC, E. P. KATZ AND M. J. GLIMCHER, *Biochim. Biophys. Acta*, 133 (1967) 97.
108 G. NIKIFOROUK AND M. GRUCA, in R. W. FEARNHEAD AND M. V. STACK (Eds.), *Tooth Enamel II*, John Wright, Bristol, 1971.
109 A. G. FINCHAM, in R. W. FEARNHEAD AND M. V. STACK (Eds.), *Tooth Enamel II*, John Wright, Bristol, 1971.
110 R. C. BURGESS, personal communication, 1969.

111 D. B. SCOTT, M. J. USSING, R. F. SOGNNAES AND R. W. G. WYCKOFF, *J. Dental Res.*, 31 (1952) 74.
112 D. B. SCOTT, *Ann. N. Y. Acad. Sci.*, 60 (1955) 541.
113 D. B. SCOTT AND M. U. NYLEN, *Ann. N. Y. Acad. Sci.*, 85 (1960) 113.
114 A. BOYDE, *Brit. Dental J.*, 121 (1966) 85.
115 R. W. FEARNHEAD, *Arch. Oral Biol.*, Spec. Suppl. (Proc. ORCA Congress Paris 1962), (1963) 257.
116 R. W. FEARNHEAD, in M. V. STACK AND R. W. FEARNHEAD (Eds.), *Tooth Enamel*, John Wright, Bristol, 1965, p. 127.
117 B. SUNDSTRÖM AND T. ZELANDER, *Odontol. Rev.*, 19 (1968) 249.
118 M. J. GLIMCHER, L. BONAR AND E. J. DANIEL, *J. Mol. Biol.*, 3 (1961) 541.
119 W. T. ASTBURY, S. DICKINSON AND K. BAILEY, *Biochem. J.*, 29 (1935) 2351.
120 K. M. RUDALL, *Advan. Protein Chem.*, 7 (1952) 253.
121 K. M. RUDALL, in M. V. STACK AND R. W. FEARNHEAD (Eds.), *Tooth Enamel*, John Wright, Bristol, 1965, p. 133.
122 H. J. HOHLING, R. M. FRANK AND R. HARNDT, *Deut. Zahnärzteblatt.*, 4 (1963) 77.
123 L. C. BONAR, M. J. GLIMCHER AND G. L. MECHANIC, *J. Ultrastruct. Res.*, 13 (1965) 308.
124 F. G. E. PAUTARD, *Arch. Oral Biol.*, 3 (1961) 217.
125 A. G. FINCHAM, G. N. GRAHAM AND F. G. E. PAUTARD, in M. V. STACK AND R. W. FEARNHEAD (Eds.), *Tooth Enamel*, John Wright, Bristol, 1965, p. 117.
126 F. G. E. PAUTARD, in M. V. STACK AND R. W. FEARNHEAD (Eds.), *Tooth Enamel*, John Wright, Bristol, 1965, p. 136.
127 L. C. BONAR, G. L. MECHANIC AND M. J. GLIMCHER, *J. Ultrastruct. Res.*, 13 (1965) 296.
128 W. G. PERDOK AND G. GUSTAFSON, *Arch. Oral Biol.*, 4, Spec. Suppl., (1961) 70.
129 W. S. S. HWANG, E. A. TONNA AND E. P. KRONKITE, *Nature*, 193 (1962) 896.
130 R. C. GREULICH AND H. C. SLAVKIN, *The Use of Radioautography in Investigating Protein Synthesis*, Academic Press, New York, 1965.
131 M. DEAKINS AND R. L. BURT, *J. Biol. Chem.*, 156 (1944) 77.
132 H. S. M. CRABB AND A. I. DARLING, *The Pattern of Progressive Mineralisation in Human Dental Enamel*, Pergamon, Oxford, 1962.
133 E. B. BRAIN, *Brit. Dental J.*, 123 (1967) 177.
134 J. E. EASTOE AND G. E. CAMILIERI, in R. W. FEARNHEAD AND M. V. STACK (Eds.), *Tooth Enamel II*, John Wright, Bristol, 1971.
135 W. E. STARKEY, *J. Dental Res.*, 44 (1965) 1175.
136 A. BOYDE, discussion in R. W. FEARNHEAD AND M. V. STACK (Eds.), *Tooth Enamel II*, John Wright, Bristol, 1971.
137 H. ROSSER, A. BOYDE AND A. D. G. STEWART, *Arch. Oral Biol.*, 12 (1967) 431.
138 E. J. REITH, *J. Cell Biol.*, 18 (1963) 691.
139 E. J. REITH, in M. V. STACK AND R. W. FEARNHEAD (Eds.), *Tooth Enamel*, John Wright, Bristol, 1965, p. 108.
140 A. R. TENCATE, *Arch. Oral Biol.*, 8 (1963) 755.
141 E. KATCHBURIAN AND S. J. HOLT, *Nature*, 223 (1969) 1367.
142 S. SAKAMOTO AND S. SASAKI, *Arch. Oral Biol.*, 14 (1969) 987.
143 J. E. EASTOE, in R. W. FEARNHEAD AND M. V. STACK (Eds.), *Tooth Enamel II*, John Wright, Bristol, 1971.
144 M. J. GLIMCHER, P. T. LEVINE AND G. L. MECHANIC, *Biochim. Biophys. Acta*, 136 (1967) 36.
145 M. J. GLIMCHER AND S. M. KRANE, *Biochim. Biophys. Acta*, 90 (1964) 477.
146 A. I. DARLING, *Ann. Roy. Coll. Surg. Engl.*, 29 (1961) 354.
147 W. F. NEUMAN AND M. W. NEUMAN, *Chem. Rev.*, 53 (1953) 1.
148 M. J. GLIMCHER, *Clin. Orthopedics*, 61 (1968) 16.

SUBJECT INDEX

(for volumes 26A, 26B and 26C)

Abdominal springs, beetles, anatomical details, 637
— tergites, locusts, anatomical details, 636
Abequose, in O-specific polysaccharides, correlation with O-factors 4 and 8 of *Salmonella*, 150
—, in somatic antigen polysaccharides, gnb, 130, 132, 139, 147, 150, 151, 157, 172
Acacia gum, hemicellulose aldobiuronic acids, 14
Acanthocephala, chitinous structures, 604
Acanthosomes, association with developing elastin, 702, 703
—, and vesicles in elastogenesis, 703
3-Acetamido-3,6-dideoxygalactose, in *Xanthomonas* lipopolysaccharide, 142
4-Acetamido-4,6-dideoxygalactose, in biosynthesis of sugar nucleotides in bacteria, 127, 172
Acetobacter xylinum, cellulose synthesis, 30, 31
O-Acetylation on C-6 of *N*-acetylmuramic acid, *S. aureus* cell walls, 59
N-Acetyldopamine, in cuticle hardening in insects, 654
Acetylgalactosamine, in *Salmonella* polysaccharide, 147
N-Acetylgalactosamine, in glycosaminoglycans, 452, 453
N-Acetyl-D-galactosaminuronic acid, in Vi-antigen, *O*-acetylation, 191
Acetylgalactose, in *Salmonella* polysaccharide, 151
Acetylglucosamine, in chitin, 596, 597
—, complexes with α-amino acids and peptides, 612
—, in somatic and capsular antigen polysaccharide, gnb, 153, 157, 172, 199, 201

N-Acetyl-D-glucosamine, in peptidoglycans of bacterial cell wall, 54, 57–60, 106, 107
Acetylglucosamine 1-phosphate, enzymatic formation, 70
N-Acetylglucosaminyl-β-1,4-*N*-acetylmuramic acid, formation by enzymatic hydrolysis of peptidoglycans, 57
N-Acetylglucosaminyl-β-1,4-*N*,6-*O*-diacetylmuramic acid, formation by enzymatic hydrolysis of peptidoglycans, 57
N-Acetylmuramic acid, in peptidoglycans of bacterial cell wall, 54, 57–60, 67, 106, 107
endo-*N*-Acetylmuramidase, peptidoglycan degradation, 62, 63
—, splitting of polysaccharide chains of peptidoglycans into disaccharide units, 57, 58
N-Acetylmuramyl-β-1,4-*N*-acetylglucos-amine, formation by enzymatic hydrolysis of peptidoglycans, 57
N-Acetylmuramyl-L-alanine amidase, liberation of free disaccharides from peptidoglycans, 57, 58
Achlya sp., lignin and lignification, 41–46
Acidosis, metabolic, and strength and composition of avian egg shell, 288
Acrotretida, periostracum, shell, proteins, amino acid composition, 764
—, shell, structural protein, ratio of dicarboxylic to basic amino acid residues, 772
—, —, various layers, proteins, amino acid composition, 767
Actinomycin D, in proteoglycan synthesis, 452, 455
Aculeata, (stinging insects), silk production, 479

Abbreviations: EM, electron microscopy; gnb, Gram-negative bacteria.

Avian egg shell, *(continued)*
— — —, polysaccharides, 277, 283
— — —, — of membranes, 277
— — —, pores, 277
— — —, protein fractions, 280, 281
— — —, protein–polysaccharide
complexes, in calcified regions, 284
— — —, resorption by developing chick
embryo, 279
— — —, sialic acid in, 283
— — —, staining reactions, in emu
shell, 285
— — —, strength, and composition, 288
— — —, sulphated polysaccharides in,
283
— — —, taxonomy, relation to
composition, 286–288
— — —, thickness relation to total N,
287
— — —, ultrastructure, 273–277
— — —, uronic acid in, 283, 284, 286–
288
Azotobacter agilis, capsular polysaccharide,
sugar constituents, 206
— *vinelandii*, capsular polysaccharide,
and cyst formation, 206
— —, —, sugar constituents, 205, 206
— —, slime polysaccharide, sugar
constituents, 206

BA-antigen, *see* Antigen, BA-
Bacillus anthracis, cell wall polysaccharide,
serological reactivity, 89
— *coagulans*, teichoic acid, decrease
in wall at higher temperatures, 96
— *licheniformis* 6346, lysozyme
solubilization, teichoic acid fractions
of digest, 96
— —, protein in cell wall, 84
— —, teichoic acids, 95, 96
— —, teichuronic acid, 90, 95, 96
— *megaterium*, teichoic acid, association
with cell membrane, 97
— — KM, cell wall
phosphomucopolysaccharide, 89
— —, strain M, polyglutamate capsule,
89
— —, —, polysaccharides of cell wall
and capsule, 89
— —, —, teichoic acid, intracellular, and
cell wall, 89

Bacillus (continued)
— *sphaericus*, Vi-antigen degrading
enzyme, 194
— *stearothermophilus*, teichoic acid,
complex with peptidoglycan, 97
— —, —, decrease in wall at higher
temperature, 96
— *subtilis*, linkage of ribitol teichoic
acid to peptidoglycan, 85
— —, teichoic acid, alanine elimination, 95
Bacitracin, inhibition of peptidoglycan
synthetase, 76
—, peptidoglycan synthesis, inhibition, 80
Bacteria, classification, and somatic O-
and K-antigens, 123
—, Gram-negative, capsular antigens,
definition, 106
—, —, extracellular polysaccharides,
definition, 106
—, —, peptidoglycans, *see* Peptidoglycans,
gnb
—, —, somatic and capsular antigens, 105–
211
—, —, somatic O-antigens, *see* Antigen(s),
O-
Bacterial cell walls, 53–99
—, biology, 53–56
—, and capsular material, 81
—, definition, 53, 81
—, gnb, proteins extractable by detergents,
107
—, —, structure, general features, 106, 107
—, Gram(+) bacteria, protein antigens,
83, 84
—, isolation and purification, 54
—, non-peptidoglycan components,
amount, 54, 55
—, —, functions, 55
—, —, in Gram(+) bacteria, 54, 55, 81–99
—, peptidoglycans, *see* Peptidoglycans,
bacterial cell wall
—, preparations, enzymes in, 81, 82
—, protein–polysaccharide–lipid
complexes, gnb layers, 54, 55, 106
—, removal of cytoplasmic and cell
membrane components by nucleases
and trypsin, 81
—, structure, influence of growth
conditions, and phase of growth, 82
—, synthesis, inhibition by antibiotics, 66,
67

Dental enamel, *(continued)*
—, maturation, 792–794
—, —, mechanism built into structure of secreted enamel, for crystal growth and protein removal, 826
—, —, protein removal, 826–829
—, —, spatial pattern, tracing by microradiography of thin sections, 825
—, —, sudden transition, 825, 826
—, —, tracing by partial demineralization by acetic acid buffer and staining, 825
—, mature, chemical balance sheet, (table), 796–798
—, —, content of organic substances, besides proteins, 797
—, —, demineralized, δ-(dental)-keratin, 823
—, —, middle and outer enamel, proteinaceous material, 809–811
—, —, organic constituents, carbohydrate content, 811
—, —, organic matter, tight binding to the crystals and protection of enamel against disease processes, 830
—, —, protein content, 808
—, —, protein and eukeratin, 809
—, —, protein isolation, contamination with collagen, 809
—, —, soluble organic material, high voltage electrophoresis, 815
—, —, 'spaces', porosity, 789, 791, 813
—, —, volume percentages of inorganic and organic substances and water, 798
—, mineralization, 792–794
—, organic and inorganic phase, relation, 829, 830
—, organic matter and water, loss during maturation, 792, 794, 795
—, organic phase, 791–798, 805–830
—, —, amorphous character, 821, 822
—, —, EM, filaments and granules, fibrous structure or gel structure, 821, 822
—, pellicle, bacterial remnants, 813
—, —, salivary glycoproteins, 813
—, plaques, composition, 813
—, post-eruptive integument, sub-surface cuticle, surface cuticle and pellicle, 813
—, prismatic, mineralizing surface, (fig.), 790
—, proteinaceous, low mol. wt., soluble material, 796, 797

Dental enamel, *(continued)*
—, proteins, gel electrophoresis, non-covalent interactions in aggregation, 821
—, —, separation, 795
—, reptiles, EM and optical microscopical structure, 789
—, rods, (prisms), *see* Enamel prisms
—, secretion by ameloblasts, 785, 791
—, stability and size of crystals, 801, 802
—, structure, features, 788–791
—, survival in unchanged form after death, 785
—, texture, 790, 803
—, water flow through intact —, 801
—, water in, 795–798
—, young, bovine, 819, 820
—, —, electrophoretic separation of neutral soluble proteins, 814, 815
—, —, gel filtration of soluble proteins, 815–817
—, —, gel preparation from amelogenin component, 822
—, —, organic phase, incorporation of labelled amino acids, 824
—, —, —, mutability, 824
—, —, —, X-ray diffraction after demineralization, cross-β structure, 822, 823
—, —, proteinaceous material, differences in amino acid composition with collagen, keratins and other proteins, 805–808
—, —, proteins, fractions separated, amino acid composition, (table), 817
—, —, —, heterogeneity, 815–821
—, —, —, and proteinaceous material, *see also* Amelogenins
—, —, —, solubility in phosphoric acid, fractionation with electrolytes, 814
—, —, —, water and neutral soluble fractions, changes during maturation, 814
—, —, transport of inorganic ions from the enamel surface to growing crystals, role of proteins and amino acids, 824
Dentalium octangulatum, shell, crystal axes, orientation, 250
Dentine–enamel junction, 786, 813
6-Deoxyhexoses, in O-specific polysaccharides, 124

Methionine, in collagens, 337
— peptides, collagen antigenicity, enhancement, 401
4-Methoxyglucuronic acid, in capsular polysaccharide of *Rhizobium*, 205
N-Methylbenzothiazolone, histochemical staining of elastin and collagen, 684
cis-11,12-Methylenehexadecanoic acid, in lipid-B of O-antigen, 114
Methyl ester group introduction, pectic substances, in growing cells, 27
Metridium, body wall, collagen, amino acid analysis, 305
Mice, elastin, 700–702
Micrococcus lysodeikticus, cell wall, peptidoglycans, 65, 73–76
— —, glucose-2-acetamido-2-deoxymannuronic acid polymer, in cell wall, 90
— —, peptidoglycan synthesis, phospholipid cycle, 74, 75
— *roseus*, cell wall peptidoglycans, 56, 57, 60–62
Microincineration for study of mineral constituents of cells, 23
Mimosa sp., lignin and lignification, 41–46
Molluscs, chitinous structures, 606, 610, 614, 620, 621, 627
—, Prosobranch, keratin-like textures of egg capsules, 586–588
Molluscan extrapalleal fluid, protein, 764
Molluscan shells, age effect on amino acid composition of periostracum, 270
—, amino acids in, taxonomic groups, (fig.), 268
—, amino acid composition, 256
—, —, and mineral phase, (table), 248
—, amino acid content, (tables), 258–265
—, —, in nacreous shell matrix, 257
—, bivalve, of mussel, anatomy, (fig.), 239, 240
—, carbohydrate in periostracum, 265
—, chitin in bivalves and gastropods, 265, 266
—, composition, environmental effects on, 268–270
—, crossed lamellar, 241, 244
—, crystal axes, orientation, (fig.), 250
—, crystal growth, and organic matrix, 247, 249
—, crystal orientation, 250–252

Molluscan shells, crystal orientation, *(continued)*
—, —, role of epitaxy, 251, 252
—, environmental effects, on composition, 268–270
—, foliate, calcitostracum, 241
—, —, crystal form, 240, 241
—, —, EM study of crystal growth in oyster, 253
—, —, nacre, 241
—, —, nacrosclerotin and nacroin, 241–243
—, formation, function of extrapallial fluid, 245, 246
—, —, inhibition by sulfonamide drugs, 247
—, —, mantle, function in Ca provision, 245, 246
—, —, organic matrix, mechanism of synthesis and deposition, 246
—, —, periodicity, day–night cycle, 253
—, —, —, mechanism of alternating lamellae, 252
—, —, permeability of mantle epithelium, for Ca, Na and K, 246
—, —, rate, 247
—, galactosamine content, 258–265
—, gastropod, protein content, (table), 255
—, glucosamine content, taxonomic groups, (fig.), 267
—, grained, 240
—, homogeneous, 240
—, layers, thickness of, environmental effects, 269
—, matrix composition, and taxonomy, 266–268
—, mineral phase and amino acid composition, (table), 248
—, mucopolysaccharides, acid, in calcified regions, 249
—, —, in calcified layers, 264
—, nacreous, 240, 242
—, organic components, water-soluble fraction, sequestering within $CaCO_3$, 254, 255
—, organic matrix, calcification, and distribution of acid amino acid residues and acid mucopoly-saccharides, 248, 249
—, —, content, 254, 255

Pachymeta flavia, silk fibroin, amino acid composition, 485
—, —, X-ray group, 485
Pachynematus foveatus, silk, cocoon, collagen structure, 528
— *kirbyi*, silk, cocoon, collagen structure, 528
— *xanthocarpus*, silk, cocoon, collagen structure, 528
Pachypasa otus, silk fibroin, amino acid composition, 485
—, —, core, amino acid content, 511, 512
—, —, X-ray group, 485
Papain, effect on protein–polysaccharides, of cartilage, effect on hydration, 440
—, elastolytic activity, non-specific, 669
—, structural breakdown of cartilage, *in vivo* and *in vitro*, 461, 462
Papilio machaon, silken girdles, 478
PAPS, *see* 3′-Phosphoadenylyl sulphate
Paraffins, in plant cell walls, 24
Paratose, in O-specific polysaccharides, correlation with O-factor 2 of *Salmonella*, 150
—, in somatic antigen polysaccharides, gnb, 130, 150
Pasteurella, R-mutants, polysaccharides, lack of galactose, 168, 171
— species, lipopolysaccharides, sugar composition, serological cross reactions with *Salmonella* groups, (table), 138, 139
— *multicida*, type I, capsular polysaccharide, sugar constituents, 207
— *pestis*, capsular polysaccharide, 207
— —, capsular protein antigen, effect on extractibility of somatic O-antigen, 206, 207
— —, surface antigens V, W, 4, 103
Pauling and Corey, collagen fibril structure, 313
Pea, lignin synthesis, experimental, 34
Peach, pectin, esterification in ripening, (table), 27
Pears, cell wall, ripening, changes in chemical composition, 28
—, hemicellulose aldobiuronic acids, 14
Pectic acid, apple, mol. wt., chain length, 11, 12
—, in model lignification systems, 39, 40
—, mol. wts., 11

Pectic acid, *(continued)*
—, structure, chain length, 12, 13
Pectic enzymes, in plant cell wall fractionation, 2
Pectic substances, in growing cells, methyl ester group introduction, 27
—, plant cell wall, analyses, 9, 10
—, —, determination, 2
—, —, solubility characteristics, 7
Pectin, apple, mol. wt., 11
—, interactions with ions, 12
—, methoxyl contents, 12
—, mol. wts., 11
—, proto-, insolubility, mol. wts., 12
—, and ripening of fruits, 27, 28
Pectinaria, glands for tube secretion, 724
—, tube-forming organs, tube construction, 721
— *koreni*, tube, carbohydrate fraction, 733, 734
— —, inorganic components, 732, 733
— —, proteins, amino acids, 734
Pectinic acids, mol. wts., chain length, 11, 12
Pelecypoda, nacreous shell, amino acid concns., 257
—, shell, glucosamine concn., 267
Pellicle, acquired, replacement of amelo-blasts after tooth eruption, 787, 813
—, dental enamel, composition, 813
Penguins, egg shells, 287
Penicillamine, α-aminoadipic acid-δ-semialdehyde, increase in elastin in chick embryo, 695
—, inhibition of desmosine formation in chick embryo, 695
—, interaction with collagen, 695
—, reaction with α-aminoadipic acid-δ-semialdehyde in elastin, 697
Penicillin(s), inhibition of transpeptidation, in peptidoglycan synthesis, (scheme), 80
—, molecular structure, similarity with D-alanyl–D-alanine end of peptidoglycan, (fig.), 79
—, peptidoglycan synthesis, inhibition of cross-linking, 78–80
Penicillin G, peptidoglycan synthesis, inhibition, 80
Pentaglycine, formation in peptidoglycan synthesis, 74

886

Protein–polysaccharide(s), cartilage, light fraction, *(continued)*
—, —, —, release from residual component, 437
—, secretion by chondrocytes, 457, 458
—, —, turnover, 460
Proteoglycan(s), *see also* Protein–polysaccharides, *and* Chondromucoproteins
—, breakdown, and lysosomal β-xylosidase, 465
—, cartilage, biosynthesis, 449–456
—, —, effect on precipitation of calcium phosphate, 443
—, —, glycoprotein component in, 437
—, — matrix, concentration, 458
—, —, monosaccharide incorporation, in embryonic cartilage, *in vitro*, 451
—, —, peptide types linking adjacent polysaccharide chains, 437
—, —, polysaccharide–protein linkage, chemical structure, 434–436
—, —, protein core, amino acid composition, 436
—, —, —, mol. wts., 436
—, —, regulation of synthesis, 455, 456
—, —, structure, 431–437
—, —, synthesis, of chondroitin sulphate–protein, (scheme), 451
—, —, —, regulation by protein synthesis, 452, 455
—, —, —, role of Golgi apparatus, 451
—, —, —, sulphation of polysaccharide component as last stage, 453
—, —, xylosylation of protein core, 451, 452
—, definition, 425
—, degradation, and mechanical properties of cartilage matrix, 439
—, epiphyseal, incorporation of polysaccharide component in metaphyseal bone, 461
—, glycosylation, first, of protein, intracellular site, 452
—, polysaccharide moieties, 432–434
—, proteolytic degradation in calcification of cartilage, 443
Proteolytic enzymes, effect on native collagen, 396–398
Proteus sp., lipopolysaccharides, sugar composition, (table), 137, 138

Protocollagen, cartilage, intracellular location, 457
—, definition, 457
—, hydroxylation by soluble enzyme from chick embryo homogenate, 457
Protopectins, insolubility, mol. wts., 12
Protoplasts, bacteria, definition, 53
Protostomia, chitin synthesis, 603
Protothaca grata, shell, amino acid concns. in matrix, 269
Protozoa, chitinous structures, 603, 604, 627
Protula intestinum, glands for tube secretion, histology, 727
Protula meilhaci, tube-forming organs, tube construction, 721
Psenulus concolor, silk, amino acid composition, 486
Pseudochitin structures, definition, 595
Pseudokeratin, dental enamel, 813
Pseudomonas aeruginosa, capsular polysaccharides, correlation between environment and composition, 203, 204
— —, —, mannuronic acid, 204
— —, lipopolysaccharide, sugar composition, 141
Psilotum sp., lignin and lignification, 41–46
Psychidae, (bagworms), silk bag, 478
—, silk fibroin, amino acid composition, 485
—, —, X-ray groups, 485
Pteria penguin, shell, crystal axes, orientation, 250
Puromycin, in proteoglycan synthesis, 451, 452, 455
Pyrocine activity of bacterial proteins, 113
Pyrophosphorylases, specific, in polysaccharide synthesis, 33
Pyrrolidine, content, in collagen, and denaturation temperature, (fig.), 322
—, —, and collagen-fold formation, 370, 371
—, contiguous, and denaturation temperature of tropocollagen, 365
— residues, and thermal stability of tropocollagens, (table), 364, 366
—, and thermal stability of collagen molecule, 333, 334, 348
—, triplets devoid of —, and melting of tropocollagens, 365
Pyruvic acid, in capsular polysaccharides of *Klebsiella*, 186, 187

Somatic bacterial agglutinogen, (BA), 177

Sorghum, SiO$_2$ content of ash, 23

Sphecidae, silk, amino acid composition, 486

Spheroplasts, bacteria, definition, 53

—, —, division, 53

—, —, reversal to normal form in culture, 54

Spherulites, crystal growth, aggregation, 238, 275, 279, 280

Spiders, orb-web spinning, different kinds of silk glands and silk, 479, 480

— silks, amino acid composition, 488, 489

— —, X-ray groups, 488, 489, 514

Spilopsyllus cuniculi, (rabbit flea), jumping mechanism and resilin, 637

Spionidae, glands for tube secretion, 722

—, tube-forming organs, tube construction, 718

Spiracle 2, in locusts, anatomical details, role of resilin in opening during flight, 638

Spirochaetopterus oculatus, glands, epidermal, secretion of chitin-like substance for tube formation, horning, 723

Spirographis spallanzanii, glands for tube secretion, azurophilic and fuchsinophilic cells, 725

— —, tube-forming organs, tube construction, 721

—, tubes, carbohydrate fraction, 737, 738

—, —, dialysable and non-dialysable components, 736

—, —, proteins, amino acids, 740–742

—, —, total mucoids, sugar composition, 739

Spongins, amino acid composition, 305, 310

Spongin A, hydroxyamino acids, content, 340

Spongin B fibres, staining, 305

Sporopollenin, plant cell wall, 25

Spruce, cellulose, mol. wt., 11, 16

Spruce wood, glycan content, 16

—, hemicellulose, mol. wt., degree of polymerization, 15

—, lignin synthesis, experimental, 34

Squid pen, chitinous structures, 597, 600, 610

Squirrel monkey, dental enamel, 807

Staphylococcus aureus, cell wall, linkage of ribitol teichoic acid, to peptidoglycan, 85

— —, peptidoglycan synthesis, phospholipid cycle, 73, 74

— —, protein antigen, responsible for agglutination, 84

— —, protein in cell wall, solubilization, 84

— cell wall peptidoglycans, 57–62, 65, 67–69, 73–76

— strains, teichoic acids, cell wall, and intracellular, structure, 92–95

— —, —, extraction, 94

Streptococci, cell wall polysaccharides, 85–87

—, group A, cell wall, polysaccharide, glucosamine release, and serological activity, 86

—, —, virulence, influence of hyaluronic acid capsules, 83

—, —, —, influence of M proteins, 83

—, —, variant polysaccharide, hydrolysis by V enzyme, 86

— D, teichoic acid, intracellular, group-specific antigen, 97

—, strain 8191, teichoic acid, association with cell membrane, 97

Streptococcus faecalis, alanyl–alanine synthetase, 69

Streptomyces aminopeptidase, degradation of peptidoglycans, 60

— *aureus*, D-cycloserine treatment, 67

— —, gentian violet treatment, 68

— —, lysine deprivation, 67

— endopeptidases, effect on peptide bridges in bacterial cell wall peptidoglycans, 64

— 32 enzyme, activity on *S. aureus* cell walls, 57

— F$_1$ enzyme, activity on *S. aureus* cell walls, 57, 60

— —, digestion of cell walls of Gram(+) bacteria, 65, 66

— SA endopeptidase, solubilization of bacterial cell walls, 60

Strophandria, silks, conformation, 544, 545

Struthio, egg shell, structure, 273

Suberic acid, from suberin, 25